I0796685

Rooted in Change

Rooted in Change

THE STORIES BEHIND SUSTAINABLE WINE

BY **Jane Masters MW**
AND **Andrew Neather**

FOREWORD BY
Sir John Hegarty

Published in 2025 by Académie du Vin Library Ltd

academieduvinlibrary.com

A CIP catalogue record for this book is available from the British Library.

ISBN 978-1-917084-70-3

Publisher: Hermione Ireland
Editorial Director: Susan Keevil
Design: brynwallsdesign
Index: Hilary Bird

Printed and bound in the United States of America

CONTENTS

FOREWORD

Sir John Hegarty

Oh god, not another book on wine. Another book talking about the 'noble grape', the meaning of terroir, its place in our culture or 'How I inherited a vineyard and found my true self'. No, please stop…

Thankfully this book isn't about anything like that. In fact, if you've never opened a bottle of wine in your life, even if you have signed the pledge, you'll find this book absorbing. Because, as you will soon see, it's about something far greater. It's about sustainability. Our need to live in harmony with our environment. Our desire to leave the world as we first knew it, in a state fit for our children and grandchildren to enjoy. And if you think that's important, then this is the book for you.

Of course, the door this book goes through is wine: its production, distribution, consumption and disposal. And this door leads to so many issues that are both important to our well-being and fundamentally fascinating. If knowledge is power, then this book is dynamite. As a vineyard owner myself, I loved it, and I learned a huge amount from its pages: it challenged my views on irrigation and provoked new thoughts about regenerative farming. But most importantly, it reminded me that sustainability is complicated. Beware anyone who says they have all the answers.

The authors – Jane Masters and Andrew Neather – help you understand the complexities, the compromises and disappointments this issue throws up in the wine world. They don't lecture; they inform. Cork versus screw cap, irrigation versus non-intervention, organic versus chemical, flight versus sea freight. Whatever the style or price of wine you prefer – indeed, whatever you consume – these issues will resonate. This book helps you navigate the contentious subjects around wine.

But what I really appreciated is that the book doesn't force its views on you. Masters and Neather let you make your own decisions. From vine to bottle to table to disposal, they cover it all... How rising temperatures are not only changing regions and their wine's profiles, but also their production. How climate change is opening up new regions. While this may be good news for some, for others it is not.

Rooted in Change doesn't shy away from difficult subjects; it reminds us that there are two sides to every argument. Thinking about health: while some may scorn drinking alcohol, the sociability, conviviality and enjoyment around wine should not be forgotten. Surely these aspects contribute to a longer meaningful life? And if you're going to stigmatize wine, then what about vending machines? Surely these should carry a health warning, not only for the sugary, unhealthy snacks they primarily dispense but also because every year some 120 people are injured by falling machines – some are killed! And what about the number of people injured crossing roads: should pavements carry a health warning? Where do we stop?

So, enjoy this thoughtful, carefully crafted, superbly researched treatise on the ways in which we can achieve sustainability through the medium of wine. I'm sure you'll love it. Of course, if you don't acknowledge that our climate is changing, or support the need for sustainability in the way we live, then this book is not for you. Nor is it for those people who think the world is flat, we're all descended from aliens, and Elvis is alive and well and living in Croydon...

INTRODUCTION
THE CHALLENGE

Looking out over Graham Martin's rows of Pinot Noir grapes marching down a hillside near Chelmsford, Essex, in the early summer sunshine, it's hard to imagine any planetary threat. Martin's Lane Vineyard grows some of England's best-quality Pinot Noir and Chardonnay grapes here in the warm, dry microclimate of Essex's Crouch Valley. But, in fact, this is the front line of climate change – and Martin, like other English growers, is a beneficiary.

The last late frost in this sheltered spot was 13 years ago. Average temperatures in southern England have risen by 2°C since the 1970s: while English weather remains a challenge – witness the very wet spring and summer of 2024 – the warming climate has indisputably boosted the English wine industry over the past 20 years.

Yet a few weeks earlier, just 300 miles to the southeast, in Chablis, France, climate change was looking far less benign. Frost struck in the last 10 days of April 2024 – the third such event in four years – forcing growers to resort to measures such as lighting thousands of candles and small stoves among the vines. Worse, the frosts came after floods at the start of April – and were then followed by massive hailstorms, with one on May 1st damaging 1,000 hectares of vines, a sixth of the area's total. Weather events such as these are becoming increasingly extreme and more frequent with climate change.

Chablis vigneron Vincent Laroche told one TV station: 'Some people have lost the whole of their harvest in a few minutes. In eight days, we've seen the full range of what's possible: frost and hail. When that falls on the plant, it cuts it in two.' Rising temperatures have gradually changed Chablis' traditionally steely character over the past 30 years, but just one month of weather extremes can wipe out a vintage.

These are just two snapshots from northern European wine regions in 2024. But recent weather events add up to a worrying pattern for the global wine industry. The effects of climate change are now impossible to ignore across every economy – but especially for this industry based on one sensitive plant, *Vitis vinifera*, the common grapevine, grown in a fairly narrow range of climate conditions in different parts of the world. In this way, wine is the 'canary in the coal mine' for climate change.

Wine has been made, drunk and traded as a valued product for millennia. Archaeological evidence shows that *Vitis vinifera* was cultivated in the Near East, in what are now Georgia and Iran, up to 8,000 years ago. Winemakers and wine companies have overcome wars, revolutions, plagues, prohibitions, diseases and pests. But those were all lesser challenges when compared to today's great predicament: climate change, caused by the increased concentration of greenhouse gases in the atmosphere, mainly derived from the burning of fossil fuels and from agriculture. What is happening in the world's wine-producing regions as a result is a prime example of the way that climate change is already affecting all of us in our daily lives.

Our climate is changing already

In 2024, France's national research institute for agriculture, INRAE, published a study of how climate change is altering wine-growing regions. It showed that if global warming exceeds +2°C, then 90 percent of coastal and low-altitude regions in southern Europe and California may no longer be able to produce good wine in economically sustainable conditions by the end of the century. And in November 2024, the International Organisation of Vine and Wine (OIV) pointed to climatic challenges across both Northern and Southern hemispheres as being major contributors to the plummeting levels of wine production in 2024, the lowest global total since 1961.

Climate change presents a serious and immediate threat to the enjoyment of millions of wine drinkers around the world – and to the livelihoods of many thousands of wine growers, workers and businesses. Some observers have predicted that it could call wine's very viability into question in places as widespread as Catalonia, Bordeaux, Chile and Australia. 'I don't know how long we can stay here making good wines, maybe 20 or 30 years, I don't know,' Miguel Torres told the UK's *The Guardian* newspaper in May 2025, speaking of his family's vineyards in Penedès, Catalonia. 'Climate change is altering all the circumstances.'

The evidence of the threat is clear. But climate change is affecting grape growing and wine styles in different ways in different places. First, average temperatures are rising – and while that may mean riper, rounder wines in northern regions like Chablis or Essex, elsewhere it's a matter of survival. In Europe, temperature records are regularly broken: in July 2022, Epernay in the 'cool' Champagne region hit 40°C and Portugal's Douro Valley registered 47°C. Warmer average temperatures increase the risk of spring frost damage because they bring forward the start of the growing season, so that frost is more likely to damage young buds. Meanwhile, too much summer heat produces grapes with higher sugar content, lower acidity levels and altered flavours – changing the flavour of wines produced.

Heat also has a human impact on vineyard workers too: in January 2023 seven farm workers died from heat stroke in the South Africa's Cape. And Stellenbosch vineyard worker Charmaine King told a journalist that in her 15 years in the job, she has noticed it getting hotter. 'Sometimes it's so difficult in the sun,' she said. 'When you come home... you are getting a headache from the sun. That didn't used to happen.'

Wine is particularly vulnerable to these shifts because of the plant that produces grapes. About 90 percent of the world's wine grapes come from varieties of the *Vitis vinifera* species which can only be cultivated within a fairly narrow range of climate conditions, in places with annual mean temperatures between 10 and 20°C. These are generally found between latitudes of 35 and 50° in the Northern Hemisphere (Europe and all of the US West Coast) and 30 and 45° in the Southern Hemisphere (the wine regions of Chile and Argentina, plus most of those in Australia, New Zealand and South Africa). Some varieties, such as Chardonnay, can be grown under a wide range of conditions – and produce a correspondingly wide range of wine styles.

But other varieties, like Pinot Noir, are extremely sensitive and struggle to produce quality fruit outside their preferred zone.

Climate change increases drought conditions too. While *Vitis vinifera* has relatively low water needs – vines' roots can go as deep as 10 metres (m) in search of water – if they don't get at least some, yields fall and grapes do not ripen. As ambient temperatures increase, vines' requirements for water increase. Many European wine regions rely on winter and spring rainfall for sufficient water reserves in the soil to sustain the vines over the summer. Without water, vines ultimately die.

In 2018, a severe three-year drought in the Western Cape saw very old, dry-farmed vines survive the first couple of years, then die, unable to survive a third dry year. Extended recent droughts in the Cape, California and Australia have seen rivers dry up and dam levels drop to dangerously low levels, so that irrigation is increasingly restricted. With such pressure on water resources there is a real question about whether grape growing will be viable in regions that rely on irrigation. 'We can't just assume we have a right to water at the expense of people,' says Louisa Rose, winemaker and head of sustainability at Australia's Hill-Smith Estates. 'We can't assume the wine industry has a right to exist in 2050 unless we improve our sustainability credentials.'

Higher summer temperatures can literally turn fiery where, after extended droughts, vegetation becomes a tinderbox – any fire that starts quickly spreads out of control. Wildfires are a natural feature of a number of climates, but not on the scale seen in recent years in California and Australia. Fire raged through vineyards around Santa Rosa, California, in October 2017, killing almost two dozen people and razing thousands of houses. Thick smoke created hazardous conditions for workers and the risk of smoke-tainted wines, while several wineries burned to the ground. Poignant photos from the Signorello Estate show the well-appointed tasting room before and after the fire: after, only a charred window frame remains. Similar events have struck South West Australia and the Adelaide Hills region recently, while in July 2022, wildfires close to Bordeaux vineyards destroyed 20,000 hectares of forest.

Climate change also means more frequent extreme weather events – and spring frosts, hail and torrential rain afflict even those regions like Chablis or southern England that might otherwise see themselves as climate winners. Widespread spring frost across Europe in 2017 contributed to historically low global wine production that year. Late spring frosts cause more economic losses in Europe and North America than any other climate-related hazards. Both the frequency and severity of damage from spring frosts appears to be increasing with climate change. In 2022, hail in June hit 30,000 hectares of French vineyards in the Gironde, Languedoc and Loire, with a third of Loire producers losing up to half of their production. Giant hailstones wreaked havoc as far south as Catalonia in August that year: the largest were 12cm across and the storm left 28 people needing hospital treatment.

And lastly, while some wine drinkers might prefer a rounder Chablis or a fruitier Loire red, fundamental changes to the character of European wines, especially, threaten to upend our assumptions about what a given region's wine should taste like. This could have profound commercial as well as psychological effects: wine is intrinsically associated with place, and, indeed, strictly limited geographical areas are at the heart of most European wine regulations – they are what makes your wine, say, a Médoc and not a more humble Bordeaux Supérieur.

At the same time, climate change will also affect how wine is distributed around the world. And as its effects mount, so the pressure on land and water resources to produce food and drinking water will increase – and a non-essential product like wine will be behind those basics in the hierarchy of priorities.

In summary, climate change is affecting vine growers' ability to consistently grow high-quality grapes in commercial volumes, and it is altering the styles of wine they can produce. At the same time, it is increasing the costs and risks of growing grapes and making wine. Volatility in production increases wine's costs per litre in low-yielding years. Meanwhile the costs of offsetting risk with crop insurance for hail and frost damage is high enough to prevent many grape growers bothering with the payment.

Sustainability is about more than climate change

Climate change is the greatest challenge in the pursuit of sustainability – but there are others too. Our planet's complex interactions between land, forests, soil, oceans, air and living organisms have kept its climate relatively stable over thousands of years. But we have now reached a point where Earth's natural, self-regulating systems can no longer cope, not just with the quantity of greenhouse gases we are emitting, but with our waste and pollution too.

Since World War II, productivity increases in agriculture brought by mechanization have depended on the massive use of new generations of chemical fertilizers and pesticides. The world currently uses around 200 million tonnes of synthetic fertilizers a year, for example – a figure that has quadrupled since the 1960s. Meanwhile, the global pesticide market is worth around $130 billion a year. We now better understand the long-term effect these substances have on soils and ecosystems – and that their use is not sustainable in the long term. Some in the wine industry have made efforts to cut back, but viticulture still remains a heavy user of pesticides, especially fungicides.

Until fairly recently, soils were considered an inert substrate providing mineral nutrients and a medium to anchor crop plants in the ground: this is what several generations of winemakers and viticulturists were taught in college. But now we understand that soils are a fundamental part of living ecosystems and that there are complex relationships between a soil's microbial life, plants and larger living organisms. Agrochemicals have a profound effect on soil health and biodiversity. Over time, the use of chemicals has led to soils becoming less fertile, producing crops lower in nutritional value and less resistant to disease. These are the unforeseen, unintended consequences resulting from the use of chemical fertilizers and pesticides. Their adoption has created a vicious circle, requiring increasingly higher inputs for lower and lower returns. In addition, chemicals can contaminate groundwater and have also contributed to alarming drops in the number of birds and insects local to their application, further affecting biodiversity.

This sort of agriculture is part of a larger global economy based on an essentially linear process of extraction and cultivation of raw materials, the manufacture of products and their use, and the disposal of the waste created: extract, sell, dispose. So material resources are used and pollution and emissions generated during the production and transport of goods. It is a system that exploits finite natural resources and creates high levels of waste: more than 80 percent of the billions of tonnes of waste generated globally each year ends up in landfill. Then, it degrades, releasing methane gas and adding to greenhouse gas emissions, toxins or other pollutants and litter, on land or in the sea.

Changing this system is the goal of sustainability – though it should be said that sustainability means different things to different people. For many of the world's population, sustainability is a question of satisfying basic material needs – food, water, energy, security, healthcare and education. For others, sustainability is about access to jobs and products which improve their quality of life – such as wine. But sustainability was most famously defined by the UN in 1987 as 'meeting the needs of the present without compromising the ability of future generations to meet their own needs' – in other words, humankind living within the limits of the Earth's resources.

On this basis, our current way of living and working is not sustainable. So sustainability requires a switch from a linear process of extract, make and discard to a circular process that eliminates waste and pollution and regenerates materials. This shift to a circular economy is based on three principles – the circulation of products and materials, elimination of waste and pollution, and the regeneration of nature. Circular thinking considers waste and emissions, from the extraction of raw materials through to the end-of-life of all products. We can participate in this using life-cycle analysis to evaluate the impacts of a product on the environment during all stages of its existence. Keeping finite resources in circulation reduces the amount of waste that goes to landfill and uses them efficiently. As we shall see, waste and by-products from manufacturing processes can be used to produce alternative products, with the aim of achieving zero waste and zero pollution.

Sustainability is a holistic concept that includes environmental, social and economic variables – and we examine some of the social and economic aspects in this book as well as environmental ones. There is growing recognition of the need to make businesses more sustainable in terms of the principles under which they operate. This is social sustainability: understanding the positive and negative impacts a business has on people, including employees, customers, suppliers, neighbours, local communities and investors.

The wine industry's economic challenge

While these are all substantial challenges, they are not the only ones facing the global wine trade. The struggle to make wine more sustainable is taking place against a backdrop of the trade simultaneously tackling other major issues too. One of the most important is the drop in global wine consumption across all major wine markets – and the limited success of the trade in attracting new generations of wine drinkers.

Declining markets are nothing new. Europe saw significant declines in wine drinking from the 1960s in the biggest consumer countries – France, Spain and Italy – thanks to changes in lifestyle and urbanization. Annual French wine consumption per head plunged from 113 litres in the 1960s to 47 litres in 2021: in total, the amount of wine the French drank fell by more than half, from 5.6 billion to 2.5 billion litres a year. Similar trends occurred in Italy, Spain and Portugal. In Argentina, too, consumption has plummeted from 80–90 litres per head annually in 1980 – 'our grandmothers drank wine all day!' says Kaiken wine estate's agronomist Nicole Monteleone, in Mendoza – to 13 or 14 litres per head today.

But for European exporters, this drop in traditional markets was compensated for by lower wine production and by the emergence of new markets further afield. Initially, southern European wines found markets north in Germany, the UK, Belgium and Scandinavia. These were followed by the US, which overtook France as the world's largest wine market in 2010, and then Asia. Global consumption stabilized at around

24 billion litres, more or less in line with global production on the back of increasing exports. The proportion of global wine production exported rose from 15 to 25 percent between 1988 and 2001 and continued to increase: half of all bottles produced today are exported.

The development of non-traditional wine-drinking markets internationally means that there has never been a wider audience or appreciation of wine globally: wine drinkers have never been so diverse in nationality and culture. The paradox is that the largest destination markets – the US, the UK, Germany, China and Japan – are all now in decline. China, which saw a dramatic rise in consumption, peaked in 2017 at seven percent of world wine consumption: this has since halved. And in the US, the number of Americans drinking wine is falling. It is difficult to identify any sizeable markets remaining with the potential to take up the slack.

This has been the global wine industry's biggest concern over the past two years. Various culprits have been identified. First, as the cost of living has risen around the world, with high inflation in the price of food products and energy, wine consumption has inevitably suffered. Wine itself has become more expensive as the costs of winemaking materials, energy and labour have increased. This is one, though not the only, reason for wine's decline in a key market: it is also failing to engage the attention of younger drinkers. Analysis by consumer research body Wine Intelligence shows a long-term pattern whereby the proportion of younger consumers of legal drinking age up to 34 years who say they drink wine at least once a month has declined steadily, over a 10-year period and across a number of markets. These younger drinkers are often more drawn to craft beers, spirits and low/no-alcohol products than wine. They care less about traditional wine marketing cues around heritage and soil types than the trade often uses to promote wine.

The decline in wine consumption is also linked to a general reduction in alcohol consumption, as consumers become increasingly health conscious. No doubt this is partly influenced by the WHO's controversial 2023 statement that 'no level of alcohol consumption is safe for our health'. The 'neo-prohibitionist' anti-alcohol lobby is gaining in influence internationally. And consumers expect greater transparency: unlike food, the wine industry has not previously been required to list ingredients or nutritional information. This is changing: in Europe, under regulation (EU) 2021/2117, wines produced from 2024 onwards require mandatory labelling information to include a list of ingredients and a nutrition declaration on the bottle, or an e-label/QR code online.

Drinking too much alcohol is unhealthy and can cause well-documented social problems: how wines are promoted and sold is part of the trade's social responsibility. Yet for the large majority of wine drinkers, wine is a positive part of a healthy lifestyle when it accompanies socializing and relaxation. It has been ingrained in European culture for millennia, lying at the heart of social and religious celebrations and everyday life. Its positive impact on our quality of life and mental health is completely overlooked by anti-alcohol campaigners and regulators. Meanwhile, certain medical research indicates that wine can have a positive physical impact on our bodies too.

Wine's decline has important economic impacts. While wine represents just 0.04 percent of world GDP, most wine regions depend on its export. Wine provides valuable jobs, many in rural communities, and contributes significantly to regional economies. The wine trade in Europe is directly responsible for three million jobs, including, for example, 60,000 in Bordeaux. The Californian wine trade employs 422,000 people in California and 1.1 million directly and indirectly across the US; it delivers $170 billion to the US economy. In 2021, Chile exported 83 percent of its wine production, contributing 0.72 percent of GDP. New Zealand exports 88 percent of its wine – worth 0.62 percent of its GDP. Unless the trade can reverse the decline in wine consumption, it will inevitably hit trade and jobs.

Yet the wine trade is under economic pressure not only as a result of falling demand. It is extremely fragmented, made up of a large number of small and medium-sized businesses in a supply chain that includes growers, producers, importers, transporters, retailers and hospitality. Many of these are struggling to make a profit, with increased employment and transport costs and taxes around the world. The difficulty in recruiting staff in both general hospitality and retail across many developed countries is widely reported. The wine trade is inextricably linked to both. And wine businesses also require specialist expertise and a high level of interpersonal skill.

Meanwhile, the production and distribution of wine is highly regulated, with national and regional governments imposing a wide range of taxes. As governments look for ways to increase revenue, increases in duty rates on wine have been significant: in August 2023, for example, the British government imposed the largest increase in duty on wine in 45 years, with another hike in February 2025. Taxes on packaging such as the Extended Producer Responsibility (EPR) and labelling all equate to additional cost: they demand resources to implement and manage. CEO Steve Finlan estimated that the British Treasury's November 2024 Budget added additional costs for the UK's Wine Society of between £5 and £5.5 million – significant for an organization with a revenue of £155 million. And at the time of writing, there seems a real possibility of a full-blown international trade war, adding further tariffs to the wine trade's costs.

So the wine trade faces three main challenges: how to navigate the impact of climate change; how to address the decline in wine consumption; and how to improve its profitability to be economically sustainable in the long term.

What is sustainable wine?

Sustainable wines are those that not only taste good but are also good for the people who make them – and good for the planet too. A sustainable global wine trade should aim for net-zero carbon emissions, minimal packaging and waste, and have positive social and economic impact. But while how grapes are grown and how wines are made and packaged are decisions taken by wine companies, consumers are part of the solution too – which, if you're reading this book, probably includes you.

Consumer awareness of the environmental and social justice issues bound up with the products and services they buy has increased enormously over recent decades. Consultancy firm EY Future's international consumer index for June 2022 indicated that while affordability is consumers' first concern amid the cost-of-living crisis, they also want to live in a more environmentally friendly way. This is especially true of those aged under 35. Part of the reason why is suggested by PwC's Voice of the Consumer Survey 2024, which found that 85 percent of 20,000 consumers surveyed across 31 countries had experienced first-hand the disruptive effects of climate change in their daily lives. Meanwhile, 46 percent said they are buying more sustainable products as a way of reducing their personal impact on the environment.

By working with the grain of this trend, the wine trade stands a much better chance of attracting new consumers. It can apply circular economic thinking from vineyard to recycling bin. Winemakers can incorporate organic by-products such as vine prunings from the vineyard and grape skins left over from winemaking as compost, or they can use them as an energy source, or convert them to value-added products. And intelligent design in how they package wines, combined with closed-loop systems to reuse and recycle materials, will reduce waste and emissions.

The bad news for consumers is that there is no single fully sustainable wine – nor indeed any such consumer product. So much depends on packaging and supply chain distribution. Simple prescriptions are difficult not least because there is no one-size-fits-all solution for making wine sustainable. The most sustainable option in one market may not be so in another. 'Sustainability means wildly different things in different contexts,' says Nigel Greening, owner of New Zealand's Felton Road. 'Almost every time you find something you think is sustainable, it generally isn't. There's no simple rulebook.'

Aside from anything else, there are few sustainability standards marked on wine labels: organic and biodynamic wines are mostly labelled as such, but that doesn't mean they're necessarily sustainable. If they come in a glass bottle from a country other than your own, they almost certainly aren't. Partly for that reason, this book doesn't make recommendations about which wines to buy. They would, in any case, be out of date by the time it's on bookshelves. Rather, our aim in this book is to examine this fascinating and vital issue in all its dimensions, so that readers can understand it better.

For instance, bulk shipping wine in big tanks and then bottling it in its final market has much lower carbon emissions than shipping heavy bottles around the world. But as RJ Botha, winemaker at Kleine Zalze in South Africa's Stellenbosch, says: 'I have a big issue with shifting bottling from a country with 37 percent employment to a relatively rich country. We need to create as many jobs as possible here.' That's a social sustainability issue, and he makes an important point. Yet as Praisy Dlamini, founder of Black-owned Adama Wines in Wellington, South Africa, says: 'It's really hard for a [small] company like this. It is right to bottle here and have the jobs here – but taking on these processes can make you uncompetitive.' So there are trade-off decisions to be made between impacts on the environment and socio-economic balance. These are complicated choices to which there is often no simple, sustainable answer.

The answer for trade intermediaries and consumers is to pay more attention to the wine they're buying, and ask more questions when they buy it. We don't want to take the joy out of wine drinking; as one Spanish producer says: 'Do people want to be taking the weight of the world's problems on their shoulders when they open a bottle of wine?' Probably not. But they can still ask: why does this store not sell more organic wines? Can they recommend a Fairtrade wine? Why isn't this bottle lighter? And so on. As wine drinkers, we can encourage a shift to sustainably produced wines by paying careful attention to the wines we choose to buy. And the simple act of recycling can make a difference too: currently, about 85 percent of wine is sold in glass bottles, which, globally, mostly end up in landfill.

First and foremost, what we set out on these pages are ways through which we can fundamentally change how we look at the things we buy. Every product has a life cycle, and in a global manufacturing economy as complex as ours, those life cycles are generally more complicated than we can possibly imagine. In these pages we will pick apart the journey of your wine from grape to glass – everything from the way a farmer fertilizes their vineyard, to the conditions the people that pick the grapes live in, to the choice of stopper for the bottle, to how that bottle is transported to you. But this analysis is possible for any product in our lives, from a carton of milk, to a carrier bag, to a car. Not just possible, but necessary: making these connections will allow us to make considered choices and encourage a more sustainable society.

So at each stage in this book, we share the stories of wine people who are addressing aspects of sustainability in their businesses and are a driving force for good as they share lessons learned and best practices. We identify hot spots and key opportunities to minimize greenhouse gas emissions and shift to a circular economy.

Chapter 1 starts with the biggest challenge – climate change – and sketches the parameters of the problem. It starts with the nuts and bolts of how the carbon cycle works and how greenhouse gas emissions from human activity are driving climate change. We look in more depth at the effects of climate change in the vineyard – particularly the places which might seem to be net winners from climate change, cool regions such as England. We examine ways that winemakers are successfully reducing their emissions in the vineyard, especially via their energy consumption. And we consider the impact vineyards have in locking up carbon in plants and soils. We also set out the accepted ways of calculating the carbon emissions of businesses.

Then, in Chapter 2, we start at the beginning of the winemaking process – in the vineyard. We look at the behaviour of vines, their life cycle, and how they express the place and the soil where they are grown – their *terroir*. Then we examine ways of growing grapes and the impact of climate change on them, especially water shortages and heat stress, from Spain to Argentina. Different winemakers explain a range of approaches to the use of fertilizers, and methods for dealing with vineyard pests. We also look at the key issue of improving biodiversity, especially in one of the most exciting places for this approach, South Africa. We consider the ways in which sustainable growers are coping with water shortages. And we weigh up the sustainability of organic, biodynamic and regenerative agriculture.

The next stage of wine's journey from the vine to your glass is in the winery. In Chapter 3 we start off with the basics of how wine is made. We move on to wine microbiology; temperature control; which chemicals winemakers use – especially sulphur dioxide; which yeasts; and how and where they mature their wine. All these things have key roles in shaping the end product. Then we examine how winemakers from Greece to Chile are reducing their carbon emissions by using renewable power and by being more efficient. We seek out best practice around the world in managing waste water, capturing carbon dioxide and disposing of unwanted end products.

Chapter 4 looks at the bottling and packaging of wine, a more complicated and interesting subject than you might imagine. This is not least because glass is easily the biggest contributor to your bottle of wine's carbon footprint. Starting in the din of a glass bottle furnace in Leeds, we consider the different packaging options for wine – glass, plastic, cans and more – how they are made and their impact on carbon emissions. We examine recycling in some detail. We also look at the merits and sustainability of corks versus screw tops and other bottle closures – as well as bottle labels and capsules and the cardboard boxes in which the bottles are transported to your wine shop.

Chapter 5 opens at a huge bulk wine bottling plant in the North of England: bulk shipping is one solution to the basic problem that wine and its bottles are heavy to transport, pushing up its carbon footprint. We examine how bottles get from France, Australia or California to their destinations worldwide, via container ships. We look at the prospects for reducing carbon emissions – at sea and by road and rail. And we consider the different ways in which wine is sold and how retailers can make that operation more sustainable.

Finally, in Chapter 6, we consider the wine trade's ethical and social responsibilities. In rural agricultural areas there have been instances of the exploitation of workers, and wine is not immune to this: we look at South Africa, and also at conditions for migrant workers in France and elsewhere. Social sustainability goes beyond the vineyards. It also involves consumers, who are increasingly conscious of the environmental and social impacts of what they buy. Can they trust the wine industry? We look at the phenomenon of greenwashing. We also examine wine producers' wider responsibilities to rural communities, from the Douro Valley to Chile, and how they can help strengthen local economies through tourism in particular. And we consider what consumers want, and what wine businesses need in terms of the skills required for success, to become more truly sustainable – so that they are still around to sell wine in 100 years' time.

Chapter **ONE**

CLIMATE CHANGE

Despite the bright early-August sunshine, there is a breeze as Noelia Callejo points across her vineyard: 'Global warming is here – and we have to adapt.' At more than 850 metres above sea level, in Ribera del Duero, north-central Spain, the vines of Bodegas Félix Callejo might seem less vulnerable than some to extremes of heat. Summer nights can be cool at this altitude; the winters are bitter, with frost possible even in late April. Yet after the burning heat of 2022, the harvest began on September 7th, more than two weeks earlier than normal. Then in 2023 there was no rain from February to June.

Noelia trained in California and Argentina, her brother José Félix in Bordeaux at Pétrus, and in Chile at Torres, before they and sisters Beatriz and Cristina took on the business started by their father Félix in 1989. Today Noelia and José Félix are the winemakers. 'When I came back from South America, I realized the potential we had here,' she says. 'So we started a journey returning to our origins – going organic, reducing yields and focusing on fresher, terroir-driven wines.'

The Callejos are on the front line of climate change – in their winery as well as in their vineyards. Like almost every other wine business on the planet, they are having to adapt, and they are also doing what they can to try and stop the problem getting worse. This is a complex challenge. Climate change is caused by us burning fossil fuels, which generate greenhouse gases in the processes that produce most of our electricity, drive our vehicles and heat our homes. So even as we adapt to climate change, we are still causing it, with the forms of energy use that are woven into almost every aspect of modern life.

The Climate Action Tracker website https://climateactiontracker.org/global/cat-thermometer/ shows a +1.3°C increase in global average surface temperature for 2023 compared to pre-industrial levels. That may not sound much – but actually it's a lot.

Climate is the long-term pattern of weather measured over decades; weather is what we experience in the short term, the interaction of temperature, atmospheric pressure, wind, humidity, cloud formation, and rain. It is a complex system in constant flux: as a result, a seemingly small change in average global temperature can dramatically shift weather patterns. Warmer temperatures affect the water-holding capacity of air, atmospheric pressure and cloud formation. And because warmer air can hold more water vapour, increasing humidity, it can cause more violent rainfall when it falls – something now being experienced by grape growers from southern England to Greece to South Africa.

This is partly why climate change isn't simply causing higher temperatures and heatwaves. Additionally, it is driving more extreme weather events including more frequent and longer droughts, more extreme storms with intense rainfall, and snow events. Over the past 20 years, 90 percent of major disasters have been caused by weather-related events. The European Environment Agency reported weather- and climate-related extremes in Europe from 2021 to 2023 that caused losses of €162 billion. Economic losses have increased over time; the last three years are all in the top-five years of highest climate-related annual economic losses.

This situation is forecast to get worse – much worse. The impact of climate change will become more severe as the global average temperature increases, creating pressure on land and water resources. With a rise in average surface temperature of +1.5°C, approximately 700 million people will be living under extreme heat conditions and 75 percent of the world's coral reefs will be lost. But that's a relatively mild scenario. An increase of +2°C would lead to two billion people being exposed to extreme heat, large areas of land becoming unfit for agriculture and the demise of all the coral reefs. In a worst-case scenario, recent projections suggest a possible increase of between 3 and 4.5°C, which would make areas of the planet uninhabitable.

Back to basics: how climate and the carbon cycle work

To understand how climate change works and the part it is playing in vineyards, it helps to go back to basics – to look at the building blocks of life on Earth.

In geological terms, Earth is 4.5 billion years old. Over time, its atmosphere and climate have slowly changed and different life forms have evolved at its surface. Earth's atmosphere both sustains life and protects it against the sun's radiation. Complex interactions between land, forests, soil, oceans, air and diverse living organisms have kept the planet's climate relatively stable over hundreds of thousands of years, through self-regulating feedback mechanisms such as the carbon cycle.

Carbon is one of the most abundant elements on Earth: it is part of all living organisms, as well as in the air we breathe, in rocks and soil, in the oceans and fossil fuels. Carbon, represented by the chemical symbol C, is found in almost pure form as diamonds, graphite and coal. Combined with oxygen, it forms carbon dioxide (CO_2) or carbon monoxide (CO) gases. It is a building block in hydrocarbons, the simplest of which is methane (CH_4), also known as natural gas. Fossil fuels such as oil and diesel include larger hydrocarbon molecules containing more carbon and hydrogen atoms. Carbon is also present in carbonates such as limestone or chalk, and also in carbohydrates (sugars) and proteins. Carbon circulates between its different forms in what is known as the carbon cycle, so that an increase in one form of carbon results in a corresponding decrease in other forms.

Plants take CO_2 from the atmosphere for photosynthesis, to produce carbohydrates and grow. When this happens, carbon atoms are bound with other elements and become part of the plant – leaves, stems, woody branches, trunks and roots. Trees and forests absorb large amounts of CO_2 in this way, so that they naturally sequester – or trap – carbon in their woody parts. But plants are the base of a natural food chain, so as they are consumed by other life forms, providing energy from carbohydrates, they contribute carbon to the bodies of those insects, birds, humans and microorganisms. Then, at the end of those organisms' lives, their remains decompose, releasing CO_2 into the atmosphere and nutrients to nourish other forms of life in a biological system. This is the short-term carbon cycle, which includes plants, trees, animals and microorganisms – the biosphere.

The long-term carbon cycle occurs over an entirely different time frame. It involves physical, chemical and geological processes occurring over hundreds of millions of years – the formation of rocks, fossil fuels, volcanos and oceans. Animal shells and bones are slowly transformed into rocks, sand and minerals. Thus, during the Carboniferous Period – 300 million years ago, before the age of the dinosaurs – fossil fuels, including coal, peat, oil and natural gas, were formed from the remains of ancient plants and organisms under intense heat and pressure over millions of years. These fossil fuels are hydrocarbons – compounds containing carbon and hydrogen atoms. When we burn fossil fuels, carbon is converted into CO_2, hydrogen into water vapour (H_2O) and heat energy is released. Fossil fuel hydrocarbons are also used in the manufacture of most plastics.

Meanwhile, soil contains both organic and inorganic carbon. Inorganic carbon comes from ores, minerals and decomposed rock. Calcium carbonate is a common mineral found in soils such as chalk or limestone and many premium wine regions are known for their calcareous soils: you can see it in the whitish stones that litter vineyards in Argentina's Uco Valley, for example, or the white, chalky soils in France's Champagne and Chablis regions. By contrast, soil organic matter contains carbon bound in living organisms and in the remains of plants, animals and microorganisms at various stages of decomposition. Plants cede carbon to soil through their roots and as plant litter when they die and decompose. Foliage and microorganisms break down rapidly, while woody parts can take years. The breakdown of dead organic matter is affected by bacterial and fungal activity. The more soil life there is, the more carbon it contains. When farmers cultivate and till soils and use agrochemicals, it leads to a reduction in soil carbon and releases CO_2 into the atmosphere.

But as well as being a key part of the carbon cycle, carbon dioxide – along with the other greenhouse gases, methane and nitrous oxide

(N_2O), and water vapour – trap heat from the sun within Earth's atmosphere. The action of these gases in this way is essential: without them, the average temperature of the planet would be an inhospitable 30°C colder. The higher the level of greenhouse gases, the higher the average temperature of the atmosphere. CO_2 is the most concentrated greenhouse gas in the Earth's atmosphere, at around 0.04 percent of dry air, or around 420 parts per million (ppm). But while methane is lower in concentration today at 1.9 ppm (up from 0.9 in the 1950s), its potential for global warming is 80 times that of carbon dioxide. Cattle and sheep farming produce methane emissions, as do coal mining, oil and gas drilling, and some decomposing waste in landfill. Nitrous oxide is the most potent greenhouse gas of all, with a capacity for global warming 280 times that of carbon dioxide. Direct emissions of nitrous oxide come from the manufacture and agricultural use of nitrogen fertilizers.

The oceans absorb CO_2 from the atmosphere, which reacts with water to form carbonic acid. In this way, oceans have a buffering effect, helping to regulate weather systems and capturing 90 percent of the heat generated by global warming. But the result is warmer ocean temperatures – which in turn reduces their capacity to absorb CO_2 – rising sea levels, and more acidic seawater, which hits marine ecosystems such as coral reefs and shellfish.

Burning up: energy use and climate change

Earth's stable climate enabled humans to thrive, evolving over 150,000 years from hunter-gatherers to become settled farmers. Some 10,000 years ago, people began cultivating crops, planting wheat and growing grapes. Even earlier, some two million years ago, humans started burning wood for warmth and to cook. For thousands of years, the only fuels we used came from biomass – wood, charcoal or animal fats.

But in pre-industrial times, humans burning wood did not add significantly to the amounts of greenhouse gases in our atmosphere. For 800,000 years the concentration of CO_2 in Earth's atmosphere held steady at an average of 250 ppm, through natural buffering and feedback mechanisms. In other words, the net flow of carbon between its different forms to atmospheric CO_2 was zero.

From the 18th century, the Industrial Revolution changed all of that, as people burned coal to generate the steam that drove mills, ships and locomotives. Then, in 1882, the Edison Electric Light Station in London launched a new era by burning coal to generate electricity. At around the same time, the first oil-powered ships appeared: the oil revolution had begun. By the time the European wine industry electrified in the late 19th and early 20th centuries, the power was generated by burning oil, while petrol powered the industry's motor trucks.

The use of fossil fuel energy quickly intensified: globally, our consumption of coal, oil and gas has increased eight-fold since the 1950s. Mass production of motor cars saw an explosion of private car ownership in the early 20th century; in the century's last decades, air travel became common. As a result of soaring energy consumption, our production of greenhouse gases has skyrocketed. Global energy-related CO_2 emissions rose to their highest level ever in 2024, at 37.8 billion tonnes, according to the International Energy Agency, an increase of +0.8 percent on 2023. Atmospheric CO_2 concentration rose to 422.5 ppm in 2024, around 3 ppm higher than 2023 and 50 percent higher than pre-industrial levels. The good news is that emissions growth was lower than global GDP growth (+3.2 percent) and that in advanced economies, energy-related CO_2 emissions decreased by 1.1 percent. This has happened with the development of renewable energy sources: particularly wind and solar energy, which, with nuclear power, accounted for 50 percent of electricity generation in the EU in 2024.

Electricity and heat production, agriculture and transport together account for 70 percent of total greenhouse gas emissions. And the developed nations that benefitted most from the Industrial Revolution are responsible for most of the increase in atmospheric greenhouse gases over time: today, the G20 countries (which include the major wine markets) account for 80 percent of the world's emissions.

The effect of climate change is especially clear in Chile. Eduardo Jordán, chief winemaker at Torres in Curicó, says: 'Since 2007 here, it has been clear that the climate is changing. This January [2025] is considered the warmest on record.' There have been more heatwaves – contributing, for example, to a series of fires in 2017 and even more devastating ones in January to February 2023, where the Chilean government declared a state of emergency in three regions and there was major damage in the central-southern wine region of Maule. Jordán says that average annual rainfall early this century in Curicó used to be 750mm; it is currently 450mm and touched as little as 150mm in 2020. In the arid northern region of Limarí, where Torres has planted its northernmost vineyards, average rainfall used to be 150mm/year and now averages just 40mm. Water reserves there last year were just 10–15 percent of normal.

In fact, 2024 was the warmest year on record globally and the first calendar year with an average temperature up +1.5°C above pre-industrial levels. The UN World Meteorological Organization's State of the Global Climate 2024 report showed human-induced climate change reaching new heights, with a trail of destruction from extreme weather that took lives, demolished buildings and ravaged vital crops. More than 800,000 people were displaced and made homeless, the highest yearly number since records began in 2008. And the report lists 151 unprecedented extreme weather events in 2024, the worst figure ever recorded. The year 2024 also saw the warmest temperatures in the world's oceans and in the Arctic Circle, which is losing ice mass at a record pace. Climate change is disrupting the hydrological cycle, producing more frequent and more extreme weather, such as the catastrophic floods in Valencia, Spain, in October 2024.

The current level and pace of change in the world in terms of climate change, social change, technology and loss of biodiversity are bigger and faster than they have ever been in human history. The effects of humans are so great that scientists now speak of us having entered the geological era known as the Anthropocene Age – the era when humans are having a substantial impact on the planet. Renowned British naturalist David Attenborough has documented the human impact on biodiversity during his long lifetime: in 1937, two-thirds of habitable land was wilderness; by 2020 this had fallen to around a third. Over the same time, the human population almost quadrupled to just under eight billion (the rate of population growth is expected to slow and stabilize at around 9.7 billion by 2050). Each year, 10 million hectares of forest are destroyed to clear land for agriculture, mining and logging. This reduces Earth's capacity to absorb CO_2 from the atmosphere and especially removes the stabilizing effect of rainforests. Meanwhile, bare or cultivated soils switch from being carbon stores to emitting CO_2 – further contributing to climate change. And the products we create from fossil fuels poison the planet: plastic waste and chemicals are found in the deepest depths of the ocean, as plastic fibres in the stomachs of deep-sea fish, while there are even microplastics in our blood cells.

A changing climate in the vineyard

Climate change in vineyards is clearly visible – and dramatic. The vine's sensitivity to growing conditions means that climate change has a marked impact on wine production. Global wine production hit a 63-year low in 2024 due to adverse weather, including frosts, heavy rains and droughts. Production in the EU in 2024 was estimated at 138.3 million hectolitres, one of the lowest figures ever recorded, and lower than the previous low in 2017. The USA saw production 16 percent below the five-year average, while the Southern Hemisphere also registered historically low production figures. Chapter 2 looks at how winemakers are adapting to the new reality in the vineyard.

Warmer spring temperatures make vines more susceptible to damage from spring frosts. Late spring frosts cause more economic losses in Europe and North America than any other climate hazard. Both the frequency and severity of damage is increased with climate change. France produced one of its smallest-ever crops in 2021 due to April frost; widespread late frost across France, Italy and Spain in 2017 caused global wine production to plunge to what was at that time a historic low.

At the same time, warmer day and night-time temperatures are producing grapes that are higher in sugar, lower in acidity, and ripen sooner. As a result, wines today are higher in alcohol, lower in acidity and have flavours that are altered in nature compared to those of 20 years ago. Harvest times across Europe have got steadily earlier: Champagne house Moët & Chandon harvested as early as August just once in the 20th century (1976). This century has already seen seven August harvests there. As we've seen, in 2022, a 40°C heatwave in early June in Spain was followed by heatwaves across Europe, with new temperature records set in many areas. Epernay, in Champagne, historically one of the coolest wine-growing areas in France, hit 40°C that July, while Portugal's Douro Valley registered 47°C.

Extended periods of drought mean that soil water reserves run low, at precisely the same time that higher summer temperatures increase vines' need for water. Drought in South Africa in 2014–2018 saw water levels in dams in the Western Cape drop to unprecedented levels. The major Theewaterskloof Dam reached its lowest point, at around just 13 percent full. By 2018, Stellenbosch growers had seen their water allocations cut by 60 percent. At Ventisquero, in Chile's Maipo Valley, they relied for most of their water supplies on seven wells when the land was planted in 2001. Now, there is just one left: the rest have dried up. And at Domaine Bousquet, in Argentina's Uco Valley, where they take water from two wells, each 180 metres deep, chief winemaker Rodrigo Serrano says that they are having to dig deeper each year.

The combination of extreme heatwave temperatures and drought can also lead to wildfires. Wildfires tore across Table Mountain, close to South Africa's Constantia region, in early May 2025, scorching 3,000 hectares of land. Fires raged in Portugal's Dão region in September 2024 and devastated Chilean vineyards in 2023. In the scorching summer of 2022, fires in the Gironde close to Bordeaux's vineyards destroyed 20,000 hectares of forest. And in 2021, large wildfires hit vineyards in northern California, Oregon, Washington State and British Columbia. Over the last decade, wine producers have suffered similar calamities in Australia, Spain and Portugal. Wildfires damage infrastructure, electricity supply and air quality. They also threaten the health and safety of vineyard and winery workers breathing in smoke.

Smoke from wildfires can ruin the wine, too: grapes can absorb smoky taints, which affects quality or even causes a whole vintage to be written off. 'It's like licking an ashtray or the burned side of a cigar,' Caleb Foster, winemaker at J Bookwalter Winery in Richland, Washington, told *Wine Enthusiast* magazine in 2023. Winemakers on the US West Coast and in Australia report that even vines up to a hundred kilometres from a major fire can be affected by smoke taint – while some vines closer suffer no ill effects. Smoke taint can also be hard to detect without lab tests. 'Tasting grapes tells you absolutely nothing,' says Foster: 'Yummy-tasting grapes can be smoked.' The smoke taint shows in the wine later.

As noted in the Introduction, climate change goes beyond hotter, drier conditions. It is fundamentally a question of more extreme weather events of all types. Winemakers' most dramatic enemy in this respect is hail. Hailstorms are becoming more frequent and more extreme. In Argentina's Uco Valley, which has always been prone to hail, Rodrigo Serrano at Domaine Bousquet, shows extraordinary pictures on his phone of the aftermath of a hailstorm that hit the vineyard in November 2024, which in 20 minutes deposited a layer of hailstones 15cm deep and tore vine leaves to shreds. 'It was the biggest hailstorm in the history of the vineyard,' says Serrano. 'It's climate change, of course.'

Meanwhile, in June 2022, hail affected 30,000 hectares of French vineyards in the Gironde, Languedoc and the Loire. In August of the same year, parts of Catalonia were pummelled by hail stones 10–12cm in diameter. Even much smaller hailstones than that can cause terrible damage to vines' stems, shoots, leaves, fruit and even bark.

Vines in cold climates: is climate change ever positive?

While there are plenty of examples of Mediterranean wineries suffering from climate-change-driven heat and drought, vineyards in cooler parts of the world are experiencing some positive effects. There are small but growing wine industries in Scandinavia and in Poland, for example. But nowhere is the pattern clearer than in England.

'We only started Hundred Hills because of climate change,' admits Stephen Duckett, owner and winemaker at the Oxfordshire vineyard. Indeed, to read some press reports, you might think southern England is about to become the Loire. English wine production has increased significantly in the past decade and the quality of the (mostly sparkling) wines being produced is undeniable. But the reality is more complicated: England's cool, damp climate remains a formidable challenge – and growers also have to deal with some of the extremes hitting winemakers elsewhere.

Simon Roberts, winemaker at Ridgeview, in Sussex, has been in the business longer than most. 'When we planted back in 1995, you definitely had a clear change between the seasons: winter started in late November, whereas now it starts January, in terms of frost,' he says. 'True winter has really diminished.'

Most growers agree that the biggest impact is in the way climate change has extended the growing season. At the end of the season, an extra couple of weeks of ripening can make a crucial difference. Yet an earlier start means that late frosts hit when the vines' buds are more advanced than they would once have been.

When Ruth Simpson and her husband bought Simpson's Wine Estate's first site in 2012, near Canterbury, Kent, 10 years of weather data showed no frost incidents in the normal growing season. But, says Simpson: 'We quickly realized we couldn't be smug about frost because the start of the growing season has shifted forward into the middle of March, and there's usually a frost in mid-April.' However, she says the risk is very site specific: in their vineyards, one lower zone is vulnerable, yet their other site was untouched by frost in 2024.

'You need to be incredibly precise about site selection – from temperature to soils to drainage to air flow,' agrees Duckett, whose main vineyard is in a dry chalk valley that benefits from warmer air spilling down from the Chiltern ridge. But he warns: 'If you're somewhere vulnerable to frost, global warming isn't going to bail you out.' That means worrying nights, especially during times of warm, clear, early-spring weather where the temperature plummets after dark, as in 2020.

But climate change isn't just about the growing season and frost. 'We are definitely seeing wetter summers with much bigger volumes of rain,' says Roberts. Many areas are suffering heavier rain in more concentrated bursts, leading to some soils not absorbing run-off. Simpson says that rainfall in the last two years has been average but milder temperatures create more humidity, increasing the risk of disease, especially mildew.

Most of all, English weather remains wildly inconsistent. 'No two years are the same at all,' says Jacob Leadley, owner and winemaker at Black Chalk, in Hampshire. A low-lying site means that Black Chalk sees damaging frosts in three out of five years; in 2020, the last frost was on May 15th. 'Whether that's climate change or just being at the northern edge of what's possible in viticulture, is hard to say,' considers Leadley.

Tackling climate change: reduced emissions

Wine producers can take steps to make their operations more resilient to the effects of climate change: we look at these in Chapter 3. But all wine businesses can take action to reduce their own contribution to climate change, by cutting their greenhouse gas emissions.

Cutting emissions at the level of a single business is evidently part of a much wider international effort in that direction. The United Nations adopted its Framework Convention on Climate Change, a multilateral treaty, in 1992, following an assessment report by the Intergovernmental Panel on Climate Change (IPCC). An annual climate conference known as the Conference of the Parties (COP) has since been held to bring the convention's 198 signatories together to negotiate responses to climate change and measure progress against agreed commitments.

At the crucial COP21 Paris summit in 2015, 195 world leaders committed to capping the increase in global average surface temperature to a maximum of 2°C above pre-industrial levels – and ideally to 1.5°C. The IPCC calculates that to limit warming to +1.5°C, greenhouse gas emissions must be reduced by 45 percent compared to 2010 levels by 2030, and then halved every decade to reach net zero by 2050. Unfortunately, the world is still far from achieving that goal.

Cutting carbon emissions across the wine value chain: cleaner electricity

When Rafael de Haan set up Herència Altés in Catalonia, Spain, with his wife Núria Altés in 2016, their new winery didn't have a grid connection. 'We thought about getting a connection but our first consideration was that it would be wildly expensive,' says de Haan. 'And the second consideration was that we didn't want to be in hock to the providers.' So they installed solar panels and batteries. 'The initial calculations of the engineer about our power needs were woefully inadequate, so we've since doubled our peak power capacity,' he says. They also switched their lead batteries for lithium ones in 2023, giving them 145 kilowatt-hours (kWh) of storage capacity.

On his laptop, de Haan can see all the operating figures for the system: when he showed me his screen, the panels were generating almost 37 kW (it was cloudy outside) and the batteries were 97 percent charged. It also showed how much they were using their diesel generator, which they still need for peak periods of power use – though in the past week it had contributed less than one percent of their energy. They have also managed their power usage more closely – for example, by turning off pumps and cooling systems at night.

All this has meant that of Herència Altés's roughly 100,000 kW a year of electricity use, 63 percent is generated on site – up from 40 percent in 2021. De Haan's goal is to reduce their diesel usage to 10,000 litres a year. But while the winery has independence from the grid, it's not a perfect solution.

'I don't like to admit it, but being off the grid is not necessarily more sustainable,' he says. 'Burning fuel directly causes higher emissions than grid electricity, given the high proportion [in Spain] generated from renewables.' And while he says that the payback time on the solar panels is about three years, the panels need renewing roughly every 25 years, and batteries and generators every 10 years, whereas paying for a grid connection is a one-off investment.

Some producers manage to generate all of their electricity needs. Bodegas Félix Callejo installed solar panels on the roof of its winery in 2021: those panels now meet all the bodega's electricity needs, even in its peak-consumption period in the summer, when more power is needed to cool its wine tanks. At Tablas Creek in Paso Robles, California, generating 102 percent of its energy needs, they actually sell some electricity back to the grid.

'Renewable energy gives you autonomy,' says Josep Maria Ribas, climate change manager at major Catalan producer Torres. Torres generates 55 percent of its energy needs from solar and biomass: Ribas says that as a result they were less affected by the energy price spike of 2022–23 than others. Elsewhere in the world, installing renewables can protect winemakers against fluctuating costs and unreliable supply: South Africa had over 200 days of power cuts in 2022. Several South African wineries have freed themselves from the constraints of the country's grid by generating 100 percent of their electricity needs with on-site solar panels, such as Journey's End, in Stellenbosch. Fetzer in California and the VSPT group in Chile also use 100 percent renewable energy sources.

Ribas is also a director of International Wineries for Climate Action (IWCA – www.iwcawine.org), a group of wineries committed to reducing emissions, established in 2019 by Familia Torres in Spain and Jackson Family Wines in the US. Its members (50 in 2024) are committed to reducing their carbon emissions with a certifiable audit trail, getting to net zero by 2050 – and to generating at least 20 percent of their energy needs renewably on site. In fact, their members generate on average 33 percent of their electricity needs; some, like Herència Altés, generate much more.

The size and structure of a business will impact the level of greenhouse gas emissions it produces, but across the wine supply chain key generators of greenhouse gas emissions are packaging, transport and distribution, and we examine these in later chapters. The above examples happen to be wineries, but for any wine company, retailer or intermediary in the wine supply chain, transitioning to using electricity from renewable energy sources in this way is the single most effective way of reducing greenhouse gas emissions quickly.

Greenhouse gas emissions from electricity generation are directly proportional to the amount of electricity consumed and the energy sources the electricity is generated from. Coal-fired power stations emit more greenhouse gases than any other energy source. Oil and gas emit less compared to coal – but hundreds of times more than renewable sources such as solar and wind, hydro and geothermal. In addition, burning fossil fuels creates air pollution by releasing particulate matter (soot), mercury and other toxic gases (carbon monoxide, sulphur dioxide, nitrous oxide).

Taking the entire life cycle of solar panels into account, solar energy generation curtails 94 percent of the CO_2 emissions of coal plants and eliminates all emissions of sulphur and nitrous oxides, mercury and particulates. After the initial investment, maintenance costs are low and solar panels can last for several decades. If connected to the grid, surplus electricity generated from renewable sources can be sold, in some countries, generating income.

The technology exists in solar and wind power to generate all the world's electricity renewably without greenhouse gas emissions. In 2024, zero-carbon sources accounted for 40 percent of the world's electricity production. But while the proportion of renewable energy has increased, world consumption of electricity increased too.

Changing over an entire nation's electricity generation to renewables is a huge task that demands significant investment – and the mix of energy sources varies substantially between countries. Spain is a world leader: almost 60 percent of Spain's electricity was generated renewably in the first half of 2024, with wind farms accounting for almost a quarter of the total. The Spanish government's target, published in 2023, is for 80 percent of the country's electricity to be renewable by 2030 and 100 percent by 2050. Spain's energy-related carbon emissions have already dropped by almost a quarter since 2000. Chile is also a leader in renewable energy, with over 68 percent of its electricity generated from renewable energy sources in 2024.

Elsewhere, progress towards net zero electricity generation is patchier. In California, renewables' share is similar to Spain's, but in the US as a whole, it is just 20 percent. Over 60 percent still comes from fossil fuels. Only 22 percent of France's electricity comes from renewables, although its unusually large nuclear power industry – 69 percent of total electricity generated – means that its carbon emissions are relatively low. But China and Australia depend heavily on fossil fuels for over 80 percent of their electricity generation. And in South Africa, renewables' contribution is under 10 percent of the total.

This is not simply a question of changing over existing capacity but of increasing it too. Renewables need to cover additional demand as well as clean up existing supply to have an impact on climate change. The UK's electricity consumption is predicted to double between 2020 and 2050, partly because of the shift to electric vehicles and heat pumps. Moreover recent delays to new wind farms and other projects coming on stream in the UK provide a reminder that decarbonization is not simply a question of generating capacity. In a May 2024 report, the UK's National Energy System Operator estimated that the nation's electricity grid would need a £58 billion investment in infrastructure – pylons, cables and substations – to connect renewable sources to the grid in order to decarbonize by 2035.

Many electricity providers offer green tariffs with a guaranteed renewable source. At major Chilean producer Santa Rita, head of research and development, Andrés Ilabaca, makes the point that although self-generated solar power currently provides only 30 percent of the energy for their irrigation and bottling operations, the majority is still powered renewably given the proportion of grid power coming from wind and solar. Nevertheless, Santa Rita has a committed to reducing its carbon emissions by 54 percent by 2033. At Montes, in Chile's Colchagua Valley, 15 percent of the winery's power needs comes from solar panels on site – but the rest of its grid power consumption is certified renewable by the supplier. It's a similar pattern at Torres's winery in Curicó, Chile: up to 10 percent comes from solar panels, the rest from renewable-certified grid power.

Meanwhile, there is increased demand for renewable energy from other parts of the drinks industry too. Influential companies, including

AB InBev, Diageo and Pernod Ricard (the largest drinks companies globally) have committed to moving to 100 percent renewable energy. Increasing demand encourages providers to invest in more renewables. The International Energy Agency's 2024 World Energy Investment report found that investment in clean energy ($2 trillion, mostly covering renewable power and energy efficiency but also including grids and storage, nuclear and low-emission fuels) is now almost double the investment in fossil fuels ($1,116 billion).

Nevertheless, the transition demands substantial investment at all levels: a tightly packed, one-megawatt-capacity solar farm requires around two hectares of space and costs around €800,000. For some larger producers, even those with strong commitments to sustainability, the costs can prove a challenge: for example, in Argentina, Domaine Bousquet's Rodrigo Serrano says that financing has thus far prevented them transitioning to solar power, though in Argentina there is now an additional incentive to invest in on-site renewable energy because of the government's withdrawal of electricity subsidies.

The downside of solar is that it can only be produced during daytime when the sun is shining. If not connected to the grid, solar panels need to be combined with battery storage – as, for example, at Radford Dale, in South Africa's Elgin region, where a stack of batteries at one end of the winery, combined with their solar panels, meets all their energy needs. Likewise in 2017, venerable Australian producer Henschke installed a micro-grid, the first such system for any Australian winery. Henschke started by conducting a comprehensive energy audit with Tandem Energy, who then designed and installed a solar and battery micro-grid with diesel backup. This can increase electricity capacity during harvest-time operations when electricity needs are highest. Solar power can also be used to power irrigation pumps in the vineyards.

Sun'Agri in southern France is also innovating with the installation of solar panels above vines. These protect vines from the sun and heat while capturing sunlight and generating electricity. The panels are moveable: the angle can be controlled to provide full protection or exposure to the sun. In spring, the panels can protect against frost, providing a thermal blanket, and can offer protection against rain and hail events. They are connected to a weather station as well as to data on plant physiology. The panels are dynamic, sitting on trackers which are controlled using smart technology known as dynamic agrivoltaics. Thus, the same land is used to produce both electricity and grapes. But for the moment they are expensive, costing around €800,000 per hectare to install.

Wind turbines are another option. Wind is a more efficient power source than solar, but it is less predictable: in latitudes where grapes are grown, the sun generally shines every day, whereas the wind does not blow consistently. But in England's windy North East, Lanchester Wines' huge Greencroft bottling plant is powered by three big wind turbines generating four million kilowatts (kW) a year, in combination with a three-million-kW solar array on the roof. And in Spain, wind is the largest source of clean energy for the grid, at 22 percent.

Cutting the carbon footprint of vineyards

Tractors are widely used in vineyards to spray pesticides, apply fertilizer, trim vegetation and till soils – and tractor diesel accounts for 40 percent of vineyard emissions. Minimizing the number of tractor passes reduces those, as well as air and noise pollution. And the less a grower drives back and forth in a tractor, the less the soil is compacted: compacted soil is harder for water to get into and to drain from, and damaging to the life of soil organisms.

Electric tractors are now commercially available as a way of avoiding these emissions, produced by Monarch and Solectrac in California and SabiAgri in France. If a grower charges an electric tractor with power produced from renewable sources, the tractor generates minimal emissions (and noise). Models vary in power, whether they are two- or four-wheel drive, their size (whether they fit between rows of vines or straddle them) and the duration of their charge. Some also use driverless technology. Meanwhile, electric utility vehicles such as Polaris's Kinetic can be used for a wide range of vineyard tasks, including those performed on rough or steep terrain. Tractors are a significant investment, amortized over more than 10 years – so the switch to electric needs to be planned and budgeted for.

The next-biggest source of emissions in the vineyard is nitrogen fertilizers, which directly release nitrous oxide into the atmosphere. This greenhouse gas has a global warming potential nearly 300 times that of carbon dioxide over a 100-year timescale. In this way, nitrogen-based fertilizers account for around 17 percent of direct vineyard greenhouse gas emissions. They are also energy-intensive to produce, like all manufactured agrochemicals. As we discuss in Chapter 2, there are ways for growers to avoid such fertilizers if they focus on their soil's health and biodiversity.

Cleaner heating

At 50 percent of total energy consumption, energy for heating is the world's largest use of power. Most heating is based on natural gas and oil. We use approximately half of the heat generated for industrial processes, and half to heat space and water in buildings; a small amount is used for cooking. The most energy- and heat-intensive industries include steel using blast furnaces to extract iron ore; cement; chemicals; aluminium; and glass. This is why glass bottles represent such a large proportion of wine's carbon footprint: we look at ways to reduce that in Chapter 4.

Governments are regulating for a shift away from oil and gas through policies to phase out fossil fuel systems and promoting alternatives like heat pumps, hydrogen boilers and renewable sources. Norway, Sweden, Finland and other countries have banned the installation of new oil boilers in the past 10 years; the UK, France, Ireland and Austria will supposedly ban oil boilers in new buildings in 2025. And in the US, California and the city of Seattle prohibit the installation of all fossil fuel boilers in new buildings.

A particular focus in Europe and North America has been conversion to heat pumps, which take heat out of the air (or ground, or water) and pass it through a heat exchanger to create heat for use in buildings. Ground-source heat pumps absorb heat from the ground via a thermal transfer fluid, which flows around a loop of pipe buried outside before going through a heat exchanger into a pump linked to radiators or a hot water system. But they require outside space to dig trenches or bore holes, as well as inside for the heat pump unit. They may be more suitable in a winery than in more constricted urban and domestic environments.

Meanwhile, electric heat pumps are a lot easier to install, and three to four times more efficient than fossil-fuel boilers. California's new building energy code establishes electric heat pumps as the baseline technology. In Europe, Germany has led the way in trying to persuade homeowners to switch to heat pumps since 2022, with a goal of 500,000 heat pumps installed a year from 2024 – though the programme has faltered because of budget cuts and a shortage of suitable skilled labour to install them. Solar thermal heaters work using thermal panels to heat water stored in a cylinder: they are already common in Greece and other Mediterranean countries.

Low-carbon district energy systems use ground heat or heat from industrial waste from a central source. Only around two percent of UK homes are currently connected to district systems, but most of Copenhagen is served by them. In London's King's Cross district, for example, the King's Cross Energy Centre houses boilers and a Combined Heat and Power plant which provides heat for dozens of shops, bars and restaurants. Such systems are best suited to urban areas, so will mostly be more appropriate for wine retailing than making.

Another low-carbon heat source is biomass energy, derived from organic matter (plants and animals). Biomass can be burned to create heat directly, converted into electricity or processed into biofuels. Biomass can include crop residues, forest residues, crops grown specifically for energy use, organic municipal solid waste, and animal wastes. The most common materials are plants such as corn, soy and wood, but any organic materials, municipal solid waste, scraps from paper and wood mills, and even vine prunings can be used. A number of wine businesses use biomass, notably the cork industry: at Amorim's two manufacturing plants in Portugal, a system of extractors collects large quantities of cork dust to be used as biofuel in the factories' heating systems.

Thermal conversion heats biomass feedstocks to burn, dehydrate or stabilize them. There are different processes. During pyrolysis, biomass is heated to 200–300°C in the absence of oxygen. This produces a dark liquid known as pyrolysis oil; syngas, a combination of hydrogen and carbon monoxide; and a solid residue called biochar. Pyrolysis oil, also known as bio-oil or biocrude, can be used to generate energy or as a component in other fuels and plastics instead of petroleum. Syngas can be converted into methane and used as a substitute for natural gas, processed into biofuels or used to make chemicals and fertilizers. Biochar is similar to charcoal and can be applied to agricultural soils, sequestering carbon, increasing water retention and helping prevent the leaching away of pesticides and nutrients.

Building codes are another key area of regulation – for instance, in the amount of insulation they mandate. The goal is to increase the thermal efficiency of buildings – their ability to retain heat in winter and keep cool in summer – thereby reducing the amount of energy required to heat or cool them and the greenhouse gas emissions that would result. Energy is lost through draughts, ceilings and roofs, walls and windows, depending on the materials used. Insulation and weather proofing can improve the heating efficiency of buildings and can be retrofitted to existing buildings as well as put into new ones.

New buildings can also be designed for maximum energy efficiency using better orientation, shape and materials (*see* Chapter 3 for more detail on this). Heating, ventilation and air conditioning using programmable thermostats, timers and sensors can reduce consumption – and other kinds of heating and cooling equipment are now efficiency rated. Double glazing reduces the transfer of heat. External shading of sun-exposed walls can prevent heat gains through windows in summer. And using natural light can reduce requirement for lighting. LEED certification (standing for Leadership in Energy and Environmental Design) is the most widely recognized third-party green building certification, with several new wineries in North America and Europe certified.

The other major source of emissions for wine producers lies beyond the vineyard, in the packaging and transport of wine. We investigate these themes in Chapters 4 and 5.

Locking up carbon

As discussed in the section on the carbon cycle (*see* page 26), trees naturally sequester carbon in their woody parts. Trees can also increase the capacity of soil to hold carbon by promoting the growth of mycorrhizae, beneficial fungi which accumulate around plant roots. Vines are long-lived plants whose roots grow deep into the soil: this can also encourage biodiversity. And whereas the widespread use of agrochemicals has led to a depletion in soil carbon, methods such as regenerative farming seek to increase organic matter and restore soil carbon, as discussed in the next chapter.

The amount of carbon captured by tree planting varies a lot depending on the type of trees and where in the world they are planted – but it can make an impact. Tree planting has become a popular choice for organizations to offset their emissions (funding energy-saving projects in the developing world is the other main such option). A whole industry of offsetting organizations and businesses has sprung up. Leading UK offsetting NGO Carbon Neutral Britain, for example, claims to have planted two million trees and offset three million tonnes of carbon equivalent gases (tCO_2e) since 2020: it will offset your carbon emissions for payments from £7.75 per tCO_2e. Such offset projects can be useful, but it is important that they support things that would not otherwise happen – otherwise they are little more than 'greenwashing', a PR exercise. And ultimately, there is no way around hard decisions that change our behaviour: we need long-term solutions if we are going to halve emissions by 2030 and get to net zero by 2050.

The other type of carbon capture, given much attention by politicians in places like the UK, is the storing of CO_2, or its transformation into a stable form through chemical reaction.

Technology is being developed, for example, to capture carbon dioxide from concrete manufacture, trapping it and making it react with chemicals to convert it into a stable material, thereby removing it from the carbon cycle. However, the holy grail for believers in carbon capture and storage (CCS), direct air capture, requires large amounts of energy to suck carbon dioxide out of the air, due to the low concentration of the gas. Despite large investments – in 2024 the UK government committed £21.7 billion over 25 years to support the development of CCS – the technology remains unproven: it seems doubtful that it will operate on a meaningful scale in the short to medium term.

Business tackling climate change: reduced emissions

Globally, we need to halve greenhouse gas emissions by 2030 to 2010 levels, and reach net zero by 2050. This will require a huge collective effort. Businesses wanting to reduce their emissions and help to tackle climate change can start by focusing on reducing their energy consumption and switching to renewably sourced electricity. The goal for a business should be to be carbon neutral – that is, when the amount of CO_2 emissions released into the atmosphere from its activities is balanced by an equivalent amount being absorbed. This is also known as net zero.

It is important to understand the scale of business emissions and where they happen. That depends very much on the type of business: energy-intensive industries such as steel production are high-emitters. Some emissions will be a direct result of their own activities, others associated with products or services that they get from third parties. But all of these contribute to the carbon footprint of a given product/industry and climate change.

Calculating a business's carbon emissions is a starting point. There are a number of international standards for carbon audits – the Greenhouse Gas Protocol, PAS 2050, ISO14064 – that provide consistent methods to identify, understand and reduce emissions.

Tools have also been developed specifically for agriculture and wine – by the International Organisation of Vine and Wine

(OIV), the Fédération Internationale des Vins et Spiritueux (FIVS), International Wineries for Climate Action (IWCA) and Wines of Great Britain (WineGB). For grape growers, there are resource use calculators available such as The Farm Carbon Calculator, Cool Farm Tool and Agrecalc.

But most of these carbon calculators rely on industry averages, which have their limitations. Averages point towards areas to focus on but do not distinguish those companies that are performing better than the industry average. Using actual activity information gives a more accurate view of emissions, allows companies to see how they benchmark against peers, acknowledges those that are doing most and encourages them to do more.

A carbon audit identifies hot spots for greenhouse gas emissions and the areas to focus on to reduce them. Reducing greenhouse gas emissions takes time, as a business transforms the way it uses electricity, heating, transport and more. In some businesses, developing alternatives to emission-intensive technologies will take years – witness the efforts of the airline industry to develop less-polluting fuels.

For the wine industry, glass is the single biggest source of carbon emissions: packaging and distribution represent over half of wine's carbon footprint. Around 85 percent of wine is sold in glass bottles, which are energy-intensive to produce and heavy to transport. Wine companies can reduce greenhouse gas emissions by moving to lighter bottles, and some glass manufacturers are trying to shift to furnaces using renewable energy sources. Chapter 4 looks at glass smelting and ways wineries can reduce their carbon footprint by using lighter bottles and alternative types of packaging.

With one in every two bottles of wine sold cross border by complex distribution networks, wine racks up substantial transport miles. Mode of transport, fuel type, vehicle age, weight of goods transported and routes to market are all factors in the quantity of greenhouse gas emissions generated. Transport relies heavily on energy-dense oil products (petrol and diesel) and accounts for 30 percent of global energy consumption. We look at this in more detail in Chapter 5.

Calculating Greenhouse Gas Emissions

The starting point for any business when reducing its greenhouse gas emissions is to understand the scale of those emissions and where they happen.

Greenhouse gas emissions are categorized into three groups known as Scope 1, Scope 2 and Scope 3, depending on their source and who is responsible for them. A business or organization's carbon footprint is the sum of its Scope 1, 2 and 3 emissions. Carbon footprints can be calculated for an industry, organization, individual site or a product. But note that one company's Scope 3 emissions are another company's Scope 1. For example, the Scope 1 emissions of a glass manufacturer supplying wine bottles are the Scope 3 of the winery using them.

Scope 1

Scope 1 emissions are direct emissions generated by a business or organization. In a vineyard, this includes emissions from fuel for tractors and nitrous oxide emissions from the use of nitrogen fertilizers. Emissions from gas used for heating, fuel or transporting goods are also in Scope 1. For glass manufacturers, the gas used in furnaces to smelt glass are part of their Scope 1 emissions.

Scope 2

Scope 2 emissions are indirect emissions from the generation of electricity a business uses. These are proportionate to the amount

Wineries use large amounts of electricity for refrigeration: and, as we saw earlier, they can substantially cut their carbon footprint by switching to renewable energy. Reducing energy consumption also saves money: through building design, energy-efficient appliances with automatic timers and sleep modes, LED lighting and movement-sensitive sensors to automatically switch lights on and off when not in use. But wineries also generate CO_2 directly by alcoholic fermentation – which normally just goes into the atmosphere. As it is produced as part of a biological process by living yeast, it is known as biogenic CO_2 and is often not accounted for in carbon audits. But biogenic CO_2 has of electricity used and how it is generated. Electricity used for cooling tanks in wineries, air conditioning and lighting in offices and warehouses all fall into Scope 2 emissions. Scope 2 emissions vary according to the energy source of the electricity – so greenhouse gas emissions are relatively smaller in France, where most electricity is generated emission-free by nuclear power, than in South Africa, which relies heavily on coal-fired power stations. Emissions from irrigation pumps in the vineyard run off the grid fall into Scope 2 emissions, whereas self-generated solar power generates negligible emissions and would be Scope 1.

Scope 3

Scope 3 emissions are produced indirectly, both upstream and downstream of a business, as a result of its activity – such as purchased products, packaging, outsourced production, business travel and waste disposal. For vineyards these include emissions generated in the production of agrochemicals, those generated by growers whose grapes the winery may buy in, and by their wines' packaging and transport.

Scope 3 emissions often account for a significant proportion of a business's total emissions: for wineries, this is typically more than 80 percent of their total carbon footprint. Successfully reducing Scope 3 emissions requires collaboration between a business and its suppliers and customers.

the same climatic impact – as well as being a health and safety risk – and technology exists to capture it. We look at this in more detail in the section on cutting winery emissions in Chapter 3.

Communicating all this information to the average consumer simply looking for a bottle of wine to have with their dinner is tricky. Nevertheless, the Swedish wine monopoly, Systembolaget, is currently mapping out the carbon footprint of all the different products it sells based on activity data from growing grapes to packaging disposal, ultimately with the aim of making the information available to Swedish wine consumers by June 2026.

In addition to reducing emissions, there are opportunities for wineries to reduce their net emissions by capturing and sequestering carbon. As a decades-long-lived perennial plant, vines sequester carbon in their woody parts and roots. Vine prunings can be converted to biochar, compost or mulch. For example, environmental asset manager Beyond Zero says that calculations it has done for Denbies Wine Estate, in Surrey, in the UK, show how by capturing carbon through trees, perennial crops and other planting, the estate could be carbon negative: its carbon sequestration outweighs its emissions from fuels, processing and other sources. The problem is that such a calculation only works when considering the sites where the grapes are grown and the wine produced in isolation: factor in bottling and distribution and it looks a lot less sustainable. Cork oak forests are another large carbon sink locking up carbon for the wine industry.

To be more ambitious, a business can aim for climate neutrality – similar to carbon neutrality but including reducing all greenhouse gases, such as methane and nitrous oxide, to net zero and eliminating other negative environmental impacts too. And to go one better than that, they can aim to become carbon negative – or climate positive – by removing more CO_2 from the atmosphere than they emit.

Conclusion

Climate change represents the greatest threat to humanity in general and the global wine industry in particular. It is already changing wine production and threatening its future in places. Combatting it – and above all reducing the greenhouse gas emissions that cause it – is one of the most urgent tasks of any business aiming to make itself sustainable. That will entail changes in almost every stage of the way wine is made, transported and sold. We will return to these areas in later chapters.

But adapting to and combatting climate change are not the full story of sustainability. There are a whole range of other aspects of wine production that need to change in order for it to be made genuinely sustainable, starting with the beginning of the process: growing the grapes, the subject of Chapter 2.

Key Lessons to reduce greenhouse gas emissions

- **Calculate your total carbon footprint**, to see where you can make the biggest difference in your operations: develop a plan to address Scope 1, 2 and 3 emissions.
- **Reduce your energy consumption.**
- **Switch to renewable energy sources.**
- **Move to lower-emission alternatives** if you are replacing equipment – for example, heat pumps for heating, electric tractors in the vineyards and electric road vehicles.
- **Design new buildings to be energy efficient.**

Chapter **TWO**

GROWING GRAPES

It's an idyllic scene on a sunny winter's day in the vineyards at Tablas Creek, in Paso Robles, central California. Under a bright blue sky, sheep graze a lush cover crop of grass between the vines, watched over by their flock guardians, yawning Spanish mastiffs. These large dogs specialize in protecting livestock: mountain lion attacks are a threat here. Tablas Creek currently uses around 100 sheep.

It looks bucolic but this is a cutting-edge example of the wine industry's effort to make itself more sustainable. In August 2020, Tablas Creek became the first winery in the world to be certified regenerative under the Regenerative Organic Certified (ROC) scheme. The term 'regenerative' has recently become something of a buzzword in wine circles. For viticulture (the growing of grapes), the system is still in its infancy, negotiating its relationship with organic farming especially, and the many other approaches to vineyard management. But regenerative addresses all the key issues for sustainable winemaking: soil health and the use of pesticides and fertilizers, water use, and coping with the effects of climate change – as well as challenges in the winery and beyond.

You heard it on the grapevine

While the vineyards at Tablas Creek give the impression of being the most natural country scene, they are, in fact, a man-made invention. Vineyards are a monocultural environment not found in nature. *Vitis vinifera* is a perennial woody creeper: left to its own devices, it sprawls in all directions, making its way up trees and rambling over any other supports it comes across in search of the sun. It can climb up to 18 metres or more, using tendrils to attach itself and producing a dense canopy of vegetative shoots and leaves as it grows – and very few, small bunches of grapes.

Humans have harnessed and cultivated this species for over 3,000 years to grow grapes to make wine, developing various methods to control the vines' vigour and to produce an economically sustainable crop. Trellis and training systems, pruning, trimming, de-suckering (removing non-fruitful shoots), shoot thinning and spraying against disease are all part of the way that we have tamed this Mediterranean creeper over time, so it produces an annual crop of ripe, flavoursome grapes suitable for winemaking.

According to the International Organisation of Vine and Wine (OIV), there were a total of 7.1 million hectares of vineyards planted worldwide in 2024 (including those producing table grapes and raisins, and young vines not yet in production). They represent a small proportion of the land cultivated for crops – about half of one percent of the total on Earth. The same issues of greenhouse gas emissions, impact on water resources, use of agrochemicals and tensions between biodiversity and viticulture exist for growing grapes as for other forms of agriculture.

Yet vineyards have some advantages over other crops. Vines have limited nutritional and water needs. They can be cultivated on land unsuitable for anything else: in Portugal's Douro Valley, for example, nothing except vines can be grown on its dry, steep slopes. And vines can live for hundreds of years: grape quality generally improves as vines get older, producing wines with greater depth and complexity. Wines from particularly old vines include Penfold's Kalimna Block 42 in Australia's Barossa Valley, where four hectares were planted in the 1880s. Bollinger's Les Chaudes Terres and Le Clos Saint-Jacques plots in Champagne were also planted in the late 19th century.

As a decades-long-lived perennial plant, vines are part of an ecosystem that can encourage biodiversity both above and below ground. They can provide a habitat for all manner of life, while areas surrounding vineyards can encourage biodiversity too, as they become habitats for insects and animals. Hedges and trees around vineyards also create shade and help to moderate temperatures. Complex landscapes where vineyards are surrounded by semi-natural habitats are less susceptible to pests and diseases. Meanwhile, vineyards sequester carbon in their woody parts and roots, making a positive contribution to the fight against climate change. These opportunities are not available in annual cropping systems such as those based on cereals.

Vitis vinifera vines are best suited to a fairly narrow range of climate conditions: wine regions have annual mean temperatures of between 10 and 20°C. Classic wine-producing regions were originally established in temperate and Mediterranean climates, as found in Bordeaux and Tuscany, although vines have since flourished with irrigation in both cool, dry climates (Marlborough, New Zealand, and Oregon in the US) and in a wider range of hot and dry conditions such as those in the Barossa Valley, Australia and the Napa Valley, California.

Vines have an annual growth cycle that includes a dormant period in winter. While they are relatively undemanding, they nevertheless do need water, sunlight, heat and nutrients when they are at their most productive in the growing season – and they are more vulnerable to weather conditions during this time. They are also particularly susceptible to various fungal diseases, such as various types of mildew.

Most cultivated *Vitis vinifera* vines are grafted onto an American rootstock to avoid the threat of phylloxera, a microscopic insect that lives in soil. Phylloxera feeds on the roots of *Vitis vinifera*, causing damage that affects growth, ultimately killing the vine. Phylloxera devastated European vineyards in the late 19th century when it arrived from North America with collectors' cuttings of Native American *Vitis* species, until the solution of grafting was discovered. Grafting involves joining together the tissues of two plants, so they continue to grow as one plant. This can be done in a nursery ('bench grafting') or on to an existing vine ('field grafting'), usually during winter. Growers graft a *Vitis vinifera* scion (the upper part of the joined plant, with the trunk, leaves and grapes) onto American rootstocks (the lower part of the grafted plant) which are naturally more resistant to the phylloxera louse. In this way, the grafted vines are protected against phylloxera but keep the desirable qualities of the particular wine grape.

Though the *Vitis vinifera* species has over 10,000 varieties, only a handful are commercially significant. Vines vary a lot according to variety: each has its own characteristics and tolerances to heat, drought and disease. When it comes to climate, as mentioned, Chardonnay is much more flexible than Pinot Noir and can be grown under a wide range of conditions – producing a correspondingly wider range of wine styles. Varieties like Pinot Noir – or Nebbiolo, or Riesling – have narrower optimum growing conditions, making them more susceptible to climate change. But, of course, the same grape variety grown in different climates produces different wine: an Australian Margaret River Cabernet Sauvignon does not taste the same as a Cabernet Sauvignon from Bordeaux. This also means that as climate changes, wines from a given region risk losing the typicity for which they are prized.

Wine grapes are smaller, rounder and thicker-skinned than table grapes and contain several hundred compounds that contribute to the flavour and quality of wine. Their sweet, acidic juice, containing nitrogen compounds, minerals and pectins, is fermented to become wine. Grape skins contain aromatic and aroma precursor compounds typical of the variety, and in the case of black grapes, polyphenols – anthocyanin pigments and tannins – giving colour, structure and longevity to red wines. Grape seeds also contain large amounts of tannin. Growers get the best wine grapes when physiological, aromatic and phenolic ripeness converge, so that sugar, acidity, flavour and tannin levels are in balance. In Bordeaux, for example, this is usually in early September.

A year in the life of vines: the growth cycle

Grapes are the product of vine phenology, the plant's annual growth cycle. This has several key stages: budburst (known in French as *débourrement*); flowering (*floraison*); *veraison*, when the grapes change colour; grape ripening, and maturity. The timing and interval between each of these stages varies according to the grape variety in question and is strongly affected by temperature. Weather conditions during budbreak, flowering and fruit set (when the flowers turn into grapes) all have a direct impact on crop yield.

In winter, vines are dormant. During this time, they can survive temperatures down to -20–25°C, assuming they have acclimatized during slow temperature decline at the onset of winter rather than extreme, rapid drops. The latent buds that will produce the following year's growth are resistant to low temperatures but start to lose their cold hardiness in spring, as temperatures warm. The vine comes out of dormancy when daytime temperatures rise over 10°C and daylight hours increase. As temperatures rise, sap moves up from the vines' roots and the buds swell, leading to budbreak in spring and the emergence of shoots and leaves.

Climate change presents real risks here. Warm winter temperatures can encourage budbreak to happen too early – which is dangerous because buds and delicate green shoots can be destroyed by spring frosts. Janine Bunker, of Danbury Ridge Wine Estate in Essex, in the UK, says that, for English growers at least, this is the danger of climate change: 'The frost days aren't changing but the growing season is extending, so that if there's a later frost, it hits the buds.'

In Bordeaux in 2023, a warm year, budbreak occurred a little early, from late March. Flowering starts when average daily temperatures are between 15 and 20°C, and, in 2023, this happened from the last week in May. Weather conditions at this point are crucial as they affect the size of the harvest. The tiny flowers are what will become bunches of grapes, and their exposure to heat, wind or water stress can damage 'fruit set' and reduce yields. But Bordeaux in 2023 was blessed with trouble-free fruit set in June, heralding a large crop.

Following fruit set, the grapes grow to become small, firm, green berries that look like peas. They are high in acid, mainly due to the presence of tartaric and malic acids, and low in sugar content. But at *veraison* – which began during the second half of July, in Bordeaux in 2023 – the grapes soften and change colour.

As grapes ripen, sugar produced by photosynthesis in the vines' leaves is transported to the grapes, where glucose and fructose sugars accumulate. In Bordeaux in 2023, this process was aided by warm temperatures in late August and the first week of September. Photosynthesis during the day is optimized between 25 and 35°C; if the air temperature exceeds 35°C, the rate of photosynthesis drops. And above a certain temperature – how high depends on the grape variety – photosynthesis and the accumulation of sugars stops altogether. Meanwhile, the respiration of malic acid in grapes, which reduces the grapes' acidity, occurs day and night. The rate of respiration increases as ambient temperature increases, so that warmer night temperatures lead to lower acidity, changing the sugar-to-acid ratio.

As the external temperature rises, so a vine's need for water for transpiration increases too. Transpiration is the vine's cooling mechanism: taking water in through its roots and up through the plant to evaporate from the leaves' stomata, the microscopic pores on the underside of the leaf. If transpiration slows, leaf temperatures rise: essentially, the plant's cooling system is slowing down. This can be damaging: the vital metabolic processes that produce sugar, reduce acid and accumulate aromatic compounds and polyphenols in the plant are all sensitive to temperature fluctuations. Heat stress occurs when the temperature rises above the plant's tolerance threshold: it is often accompanied by water stress from drought conditions.

Increasingly, these processes can be measured by technology. Joe Uhr, winemaker at Gundlach Bundschu in California's Sonoma Valley, says: 'Sensors are so cheap now, and you can hook them up to cloud-based systems.' He explains that they use not only soil moisture probes but also Tule moisture sensors, which sit above the canopy measuring evapotranspiration rates. An algorithm then tells him how much water the vines are metabolizing. Indeed, Eduardo Jordán, chief winemaker at Torres in Curicó, Chile, gets a live feed on his phone from his vines' water sensors.

After harvest, the vines continue their life cycle. Their leaves continue to photosynthesize and produce carbohydrate reserves that collect in the roots. This sustains the plant during its dormancy in winter – and provides the necessary energy for budbreak and flowering the following year. Shorter days and cooler autumn temperatures stimulate the vine to prepare for winter. Latent buds form at the base of the leaf which contain primordial shoots for the following year's growth. Leaves change colour and fall, and the vegetative shoots of the current year's growth become lignified – rigid and woody. The fallen leaves release carbon into the short carbon cycle as they are decomposed by soil microbes, while the carbon in the vine's woody parts can be stored for decades. The vine then waits for another spring, so the cycle can begin all over again.

Terroir

To make excellent wine, farmers work with great attention to detail in the vineyard, controlling vine vigour and producing a decent, fully ripe crop. But many winemakers, especially in Europe, would argue that it all comes down to terroir, the hard-to-translate French term that means the combination of climate, soil, geology, altitude and aspect (the direction it faces) of a particular site. Terroir defines a site's potential for producing a particular quality and style of wine, and, over time, vine growers have matched the most suitable wine grape varieties to different terroirs. But in any given year, weather conditions influence the quality and quantity of grapes produced – making good and less-good vintages.

Differences in terroir might seem obvious when you compare, say, the frost-prone slopes of Burgundy and the broiling hot Barossa Valley in Australia. Where it gets complicated is in the small but significant variations within small areas. Within regions or even within a single vineyard, differences in terroir result in a wide range of wines.

Even a short walk in many Burgundy vineyards – and a tasting of their wines – illustrates the point. In the classic white Burgundy appellation around the small village of Meursault, for example, there are 19 different areas designated as 'Premier Cru' – a step up in quality from the wines labelled with just the name of the village. Stand on the edge of a vineyard in the Les Perrières Premier Cru and take just a few steps across a single-track road and you're in Les Genevrières Premier Cru, which produces a wine that tastes clearly different. And yet 400 metres down the hill, the Chardonnay grapes qualify only for humble AOC Bourgogne – not even Meursault at all. There are significant financial implications to this: currently, up to around £100 (US $125) a bottle difference in price between the Premier Cru wines on the hill and those at the bottom.

You can see myriad differences of terroir within other vineyards too. Sitting in recently planted vineyards in Gualtallary, in the Uco Valley, in Argentina, Zuccardi Wines' chief viticulturist, Martin di Stefano, gives a brilliant exposition of the terroir. Here, he says, the three main factors influencing different sites' terroir are altitude, soil and the distance to the mountains of the Andes' Cordillera Frontal looming in the distance. The soils are a patchwork, and harvesting times vary hugely as a result of highly localized differences in temperature: he gives the example of a month's difference in picking time between two different Chardonnay sites here just eight kilometres apart.

The difference between the Uco and Burgundy, however, is that Argentina doesn't have a highly prescriptive appellation system: 'We don't like rules – this is Argentina!' jokes di Stefano. Yet the world's most prestigious wines generally do come from very specific vineyard sites: terroir is the basis of appellation systems used in many wine-producing countries, especially most of those in Europe. This is one of the reasons why even relatively small changes in climate are already changing the typical characteristics of wines from given appellations or vineyards. And that will have huge implications for those classification systems – and for the wine industry.

Soil

Of all the components of terroir, soil is generally seen as the most crucial. Wine regions around the world have put a great deal of effort into the detailed analysis and mapping of soil types. Appellation boundaries are often drawn according to soil and subsoils, so Chablis is known for its Kimmeridgian clay, Australia's Coonawarra for its iron-rich, red Terra Rossa clay, and Chianti for *galestro* and *alberese* soils. Soil anchors the vine to the ground and provides water and nutrients to vines through their root systems, which can go deep down for two to three metres or more until they hit solid bedrock.

Soils are characterized by the proportions of sand, silt and clay particles, organic matter and by the number of larger pebbles, stones and rocks they include in their mix. These components define the structure of soil, which affects fertility, water-holding capacity and drainage. A key characteristic of any soil for growing vines is its capacity to hold water and to release it as required. The amounts of nitrogen, phosphorus, potassium, calcium, as well as other macro and micronutrients that a vine requires, are affected by a soil's water-holding capacity, pH, organic content, minerals and plant-assimilable nutrients.

But soils are also complex ecosystems, a phenomenon the complexity of which we are still discovering. As Martin Kaiser, viticulturist at Doña Paula in Mendoza, Argentina, says: 'I was taught when I was a student that the soils here were an inert sponge that you had to add everything to.' Such assumptions were common in agricultural colleges until relatively recently. But organic matter in soils is vital. It consists of living organisms and the remains of plants, animals and microorganisms at various stages of decomposition. Organic residues such as foliage are broken down by soil microorganisms, releasing nutrients and carbon back into the carbon cycle. Fungal hyphae in the soil create branched networks of filaments that bind soil particles together to create water-stable aggregates which create pore space in the soil, enhancing water retention and drainage.

Soil life matters because soil microbes convert organic matter into nutrient forms the vine can absorb. Nitrogen, for example, is abundant in most viticultural soils, but mostly as organic material from the remains of plants, animals and microorganisms which is unassimilable by plants. Soil bacteria slowly break down organic matter into forms that can be absorbed by vines – a process known as mineralization. Humus, dark organic material that results from decomposed plant and animal matter, increases water retention in soil; it also improves vine nutrition as it absorbs and retains minerals in soil. It stimulates plant growth, aids the absorption of phosphorus and the development of microbes. It has a buffer effect regulating soil pH, absorbing heat and moderating temperature changes.

Meanwhile, mycorrhizal fungi grow symbiotically with vines' roots, providing the vine with nutrients and extending the absorptive area of its root system in exchange for photosynthesized sugars, which the vine supplies to the fungus. Larger creatures, such as beetles and springtails (the latter only 6mm long), can have a significant effect on a soil's porosity and the spread of beneficial fungi. Worms loosen, mix and oxygenate soil and their castings are rich in nutrients. So a healthy soil isn't just a question of chemical content and water. It is a living ecosystem with complex interactions between plants, insects and arthropods, earthworms and nematode worms, and millions of different microbes. It is estimated that nearly a quarter of the planet's biodiversity lives in soil, and that soils contains 50–70 percent of the organic matter on Earth.

But soil life is depleted by chemical fertilizers and pesticides. And the tilling of soil disrupts fungal networks, and at the same time releases carbon dioxide into the atmosphere. Meanwhile, the repetitive passage of tractors compacts soil, which has a negative impact on the life within it too. The importance of this ecosystem for growing grapes was first highlighted by French microbiologists Claude and Lydia Bourguignon. At a conference in 1992, they famously argued that many soils in Burgundy were so exhausted from the long-term application of agrochemicals that they had no more life to them than those of the Sahara – in other words, that conventional methods of viticulture were unsustainable. They were attacked as alarmists by some but have nevertheless attracted many followers – and clients for their consulting business, in the shape of iconic French estates that include Burgundy's Domaine de la Romanée-Conti and Saint-Emilion's Château Ausone. It is now more widely accepted that the use of agrochemicals and tilling has led to a general reduction in soil life and carbon.

Cultivating grapes

How vines are farmed impacts the ecosystem of which they are part, as well as the greenhouse gases emitted. The look of a vineyard can tell you a lot about the farming methods used: neat, tidy, disease-free vineyards often rely on the use of herbicides, insecticides, fungicides and chemical fertilizers. Less conventionally farmed vineyards – especially those managed using organic, biodynamic or regenerative viticulture – generally look less tidy, with weeds or cover crops growing between the rows and even between the vines themselves. The vast majority – over 90 percent – of the world's vineyards are farmed conventionally, with organically certified vines accounting for just six percent of the world's vineyards in 2019.

Vines are generally cultivated in rows, orientated to capture or to protect grapes from the sun. For example, Gerardo Leal, vineyard manager at Santa Rita in Chile's Maipo Valley, explains that their new plantings are aligned to avoid scorching by the hottest afternoon sun. Likewise at Montes, in Chile's Colchagua Valley, vine rows are oriented north–west to avoid too much sun. Vines can be free-standing (bush vines) or trained, but most have a trellis structure to support them. Pruning, training and trellis systems organize vines spatially, allowing exposure of leaves and fruit – and making vineyard work easier.

How many vines? Planting densities vary, from fewer than 1,000 vines per hectare to up to 10,000 in traditional regions such as Médoc, Burgundy and Champagne. European vineyards tend to cultivate higher density, lower vines, while New World regions tend to have taller structures with wider rows and more space between vines, allowing a higher degree of mechanization. Trellises vary from relatively simple wooden or steel posts and wires to large, complicated structures – and a few regions still train vines up trees.

Winter pruning, in January and February in the Northern Hemisphere, prepares the vines for the start of a new annual growth cycle. Winter pruning removes up to 85 percent of the previous year's shoot growth. That's a lot of vegetation: a vineyard with a yield of 10 tonnes of grapes produces one to two tonnes of prunings. The timing of pruning also affects budbreak: late pruning helps delay budburst and protect against spring frost. And then the vine's annual growth cycle begins as temperatures rise above 10°C.

In vineyards susceptible to frost, spring is a nerve-wracking time, as yields can be wiped out in one night: growers spend many a sleepless night monitoring temperature lows in the early hours. Climate change seems to be increasing the frequency and severity of damage from spring frost. Growers can use various methods to protect the crop. They can use frost fans to blow warmer air through a vineyard: these cost around £30,000 (US $37,500) apiece. Alternatively, large candles can create a convection current – these are perhaps a cheaper option, at about £7 (US $8.75) each, except that even a fairly small vineyard can use hundreds each winter. Or they can make smoke by burning bales of hay, both of which create pollution and damage air quality. Water aspersion, a method where growers spray the vines with water so that it freezes around the shoots to protect them, is a favoured method in Chablis. Another alternative is to attach solar-powered infrared wires to the trellising system; these raise the temperature around the shoots by about one degree: notable English sparkling producer Ridgeview, in Sussex, uses these. They work well but are expensive to install.

During the growing season, leaves and shoots grow rapidly – up to 2–3cm a day – and canes are often attached to the trellis systems to support vegetative growth and the weight of grapes. Through canopy management and crop-thinning techniques (leaf trimming, de-suckering, shoot thinning) growers achieve a balance between vegetation and fruit. Canopy management helps protect vines against fungal disease by allowing air to circulate, and exposes the leaves and bunches to the sun for effective photosynthesis and ripening. Planting density, as well as trellis and training systems, can all affect disease too. For instance, by using an open-canopy structure, which maximizes air movement and exposes foliage to sunlight and wind, a grower can reduce vineyard humidity and the risk of fungal disease, and hence the number of chemical sprays required. An open canopy also allows more effective spray treatments to reach all vine leaves and growth susceptible to infection.

Conversely, growers can allow more canopy growth to provide shade for grapes in regions where there is a risk of sun burn. After *veraison*, as grapes ripen, they become more susceptible to heat damage. Black-coloured red grapes absorb heat and can get much hotter than the surrounding air temperature. One alternative method to shield them is to use sunscreen cloth: Facundo Bonamaizon, viticulturist at biodynamic producer Chakana in Luján de Cuyo, Argentina, says that these can cut the amount of ultraviolet light reaching the grapes by 15 percent. As part of his low-intervention approach in the vineyard, Laurens Hartman at Karanika, in Amyndeon, northern Greece, uses Zeolite spray (an aluminosilicate) to reflect sunlight, kaolin as sunblock, and tea extracts such as valerian to combat heat stress.

How much water a vine needs depends upon the ambient temperature, wind and relative humidity in a vineyard. Most European wine regions depend on natural rainfall. The best terroirs regulate water supply in such a way that vines are under moderate but never extreme water stress. For example, the gravel soils that dominate in the Médoc, in Bordeaux, retain enough water reserves for the vines during dry periods, but little enough that they still have to reach out to get what they need. The Médoc was originally marshland, drained by Dutch engineers in the 18th century. The Gironde estuary keeps the subterranean water table topped up, giving the vines' deep roots access to water in summer. Drainage channels and the gravel soils evacuate excess water when heavy rains fall. Dry-grown (unirrigated) vineyards are more resilient to drought, as the vines' roots are more resourceful, descending and spreading out in search of water.

High summer temperatures, wind and low air humidity increase the amount of water the vine needs. As climate change increases average temperatures and sees record extreme temperatures, vines' water requirements increase. Up to a point, some water stress can contribute to better-quality grapes. Most of Bordeaux's most highly regarded vintages are those where there was low rainfall, producing highly concentrated, flavoursome grapes. However, higher maximum temperatures during summer and warmer night-time temperatures change the fundamental sugar-to-acid balance in grapes. Excessive hydric stress can lead to blocked grape maturation, as the vine's metabolism simply shuts down, affecting sugar and acidity levels and wine quality. Hydric stress lowers yield and decreases the vine's resistance to pests and diseases too, and can even affect the following year's crop.

And if heat and/or water stress is extreme, or lasts a long time, it can cause irreversible damage to the vines. Driving on mountain tracks to reach one of her high-altitude vineyards in Crete, Alexandra Manousakis points out a plot of dead vines: they belonged to a local smallholder but died in the extreme summer heat of 2023. Chile is suffering more than most wine producers from the effects of climate change, with prolonged drought and summer temperatures up to 40°C: January 2025 was the warmest on record. The high temperatures are contributing to the decision of some Chilean producers to rip up their vines.

This is also because Chile is a prime example of the use of irrigation in the New World, where it enables grape growing in low-rainfall regions and even in desert conditions. The arid Limarí and Elqui valleys in northern Chile and Mendoza in Argentina, fed by meltwater from the snow of the Andes, would not be cultivable without irrigation. Most of Australia's vineyards rely on irrigation from the Murray-Darling river basin and the Murrumbidgee irrigation system. The Terra Rossa soils of Coonawarra, which produce fine Cabernet Sauvignon wines, are only able to do so with irrigation, since the shallow topsoil they grow in sits over a solid horizon of calcrete rock that vine roots are unable to penetrate.

Such places, it seemed for many years, were a grower's paradise, with warm, dry climates, sunshine to limit disease pressure, and water on demand from nearby lakes, rivers and municipal systems, or dams and bore holes. In many such regions, water was cheap and taken for granted. Now, with climate change, as we saw in Chapter 1, water is becoming an urgent problem for them.

But different grape varieties react differently to heat and water stress. Theodora Rouvalis, of Rouvalis Winery in Greece's northern Peloponnese, says that in the very hot summer of 2024, 'We noticed very clearly that international grapes planted at 500–700 metres had [heat stress] problems – their maturation was too fast, and the harvest was three weeks early. But with late-ripening indigenous grapes at 700–1,000 metres – Roditis, Mavrodaphne and so on – we had no problems, and they were picked just one week early.'

In southern France, Grenache copes better than Syrah, and Cabernet Sauvignon better than Merlot. Some varieties use water more efficiently, slowing down transpiration as soon as temperatures start to rise: these are known as isohydric varieties. By contrast, anisohydric varieties keep on transpiring until the water runs out. So Grenache maintains its water supply longer than Shiraz by closing its leaves' stomata sooner in response to a soil water deficit. Other isohydric varieties include Carignan, Greek Assyrtiko and the Sicilian grapes Grillo and Cataratto: it is no accident that this happens as they are native to hotter Mediterranean climates. Most cultivated *Vitis vinifera* grapes are grafted onto American rootstocks, which themselves vary in their ability to cope with dry conditions, with 140 Ruggeri and 110 Richter the most drought-resistant rootstocks.

Temperature also affects the typical aromas of a grape variety, which derive from specific compounds. For example, thiols produce citrus and passion-fruit characters in Sauvignon Blanc, while rotundone is behind the spicy, peppery notes in Syrah and the petrol character in many Rieslings. High heat and water stress produce grapes that are smaller, dehydrated, and have fewer aroma precursors; instead, overripe, cooked-fruit characters can develop. Heat and water stress also affect the formation of the phenolic compounds, anthocyanins and tannins important for red wine colour and structure: stressed red grapes can develop hard, astringent characters. High temperatures inhibit anthocyanin synthesis, leading to paler coloured wines. Some moderate water stress helps to enhance colour, but excessive water stress negatively impacts colour intensity.

The vine is also prone to pests and diseases that reduce the quantity and/or quality of the annual harvest. *Vitis vinifera* is particularly susceptible to fungal diseases throughout the growing season, which can infect buds, leaves, shoots, grapes and the woody parts of the vine. Downy mildew, powdery mildew (aka oidium) and botrytis are all common diseases that impact the quantity and quality of grapes produced. A severe attack can reduce the crop to zero. Previously infected vineyard areas and neighbouring vineyard plots are a source of infection, making it virtually impossible to grow quality grapes without some fungicide use. An important part of all vine growers' work is spraying, throughout the growing season, to combat disease. The products growers use range from manufactured agrochemicals in conventional viticulture to naturally occurring substances used in organic and biodynamic viticulture.

Harvest time

Harvest is the moment when a year's worth of effort by the grape growers literally comes to fruition. As the main ingredient of wine, grape quality determines the quality and style of wine that can be produced. Grape quality is reflected in the balance of sugar, acidity, tannins and flavours: these are measured and assessed in the run-up to harvest in order to decide when to pick. At this moment, weather can have a high impact in a very short space of time. Rain can cause grapes to swell, becoming dilute in flavour, or, worse, splitting and allowing the fast development of rot. Heavy rainfall can make it difficult for tractors or harvesting machines to get into vineyards. So growers decide picking dates based on how grapes are maturing *and* on the weather forecast.

Warmer average temperatures as a result of climate change mean that grapes are ripening earlier and must often be picked in high summer temperatures. This has an impact on field workers as well as on how the grapes are handled in the winery. Higher ambient temperatures at picking increase the risk of spoilage and uncontrolled fermentation, resulting in lower wine quality. Hotter incoming grapes also increase the energy needed for chilling in the winery.

Grapes can be picked by hand or by machine: both methods have their advantages. Harvesting machines are cheaper and faster, so that, if rain is forecast, growers can bring the grapes in quickly. One human grape picker can pick one to two tonnes of grapes per day, whereas a machine can harvest 100–200 tonnes. Harvesting machines can also pick at night, in the dark, and in this way the grapes gain in quality as they benefit from cooler temperatures. However, machines can be used only where there are suitable trellis and training systems in place – bush vines cannot be picked by machine – and they are no use on steep vineyards like those in parts of Germany, Greece, Portugal and northern Spain.

Meanwhile, picking by hand allows for a high level of triage, where only the very best bunches of ripe, healthy grapes are selected at each pass through the vines: this is why it is preferred in most premium vineyards. And this is why Oliver Esparham, general manager of Law Estate Wines in Paso Robles, California, says: 'It's skilled work, picking. You need someone who knows what gets dropped and what goes in the box.' But finding large numbers of grape pickers at harvest time is becoming more and more difficult, as explained in Chapter 6.

Working in the vineyards

Viticulture is hard work. Winter pruning is especially hard, working in colder temperatures. 'This is skilled labour – you can't just get someone off the street and say, "go prune this",' says Hilary Graves, vineyard manager at Booker Wines, in Paso Robles, California. But during the growing season other operations, including attaching canes, de-suckering, lifting training wires and monitoring for disease, are also carried out by hand. Alex Dale, of Radford Dale in Elgin, South Africa, says experienced staff are especially important for an organic producer like him because they can identify problems in the vines, such as disease, so that he can take early preventative action.

It can be hazardous work too. During spraying operations workers can be exposed to pesticides: they need to be properly trained and protected. And in summer, vineyard temperatures can top 40°C: workers need shade, water and sunscreen. In September 2023, four pickers died of heat stroke and suspected heart attacks in Champagne, after toiling in 35°C heat. In high-density, low-trained vines, the work can be particularly backbreaking.

Fertilizers

During the growing season, the vine takes nutrients from the soil to grow and produce grapes. These nutrients are exported from the vineyard when grapes are picked and made into wine. In conventional farming, growers replenish nutrients by applying chemical fertilizers. Synthetic nitrogen fertilizers – invented early in the 20th century – transformed global agriculture, paving the way for the post-war Green Revolution and increasing crop yields to feed the world. But they have major downsides. While mineral nitrate and ammonium-based fertilizers are readily assimilated by plants, the fast release and concentration of these can have a detrimental effect on microbial life and invertebrates in the soil. Nitrate is easily leached from soils as run-off, polluting river systems, and has an impact on freshwater ecosystems too. In addition, too much nitrogen encourages excessive growth of the vine canopy, which encourages fungal diseases.

'The main problem of conventional agriculture is with fertilizers,' says leading biodynamic grower Alvaro Espinoza in Chile's Maipo Valley. 'If you use nitrogen, it makes the plant weaker. It's like sugar for humans. That's the vicious circle of these products.'

There are natural alternatives to synthetic nitrogen-based fertilizers. This is the main reason that producers who grow cover crops between their vines plant leguminous plants such as beans, which fix nitrogen in the soil, replacing what has been taken out by the grape harvest. At Radford Dale, Alex Dale sows a cover crop with this aim, choosing fava beans and lupins, which he ploughs into the soil at their highest nitrogen point. He also plants tillage radishes to improve soil structure. Meanwhile, there are also naturally derived nitrogen products for sale, such as sprays made from seaweed: Dourakis, an organic producer in western Crete, uses these.

If a grower can maintain good levels of organic matter and microbial activity in their vineyards' soils, they can further eliminate the need for chemical fertilizers to supplement soil nutrients. Alvaro Espinoza says that after more than 25 years farming biodynamically, his soils now contain four percent organic matter. Microbial life in soil raises levels of organic matter and locks in carbon: it reduces the need for chemical use, and perpetuates the further development of more microbial life, thus creating a virtuous circle, using resources more efficiently and reducing inputs and pollution.

Meanwhile, manure brings a combination of nutrients in the short and long term, as slow-decomposing humus material progressively releases nitrogen. Hence its importance for regenerative vineyards like Tablas Creek, whose flock of sheep collectively fertilize the vineyard with over 200kg of manure each day. This increases biological activity in the soil, improves soil structure, increases the soil's water retention, regulates its temperature and maintains soil health. And at the same time as fertilizing the soil, grazing animals, such as sheep, help to control weeds.

Cover crops, livestock manure and compost can all help to build soil organic matter. Dry manure, grape pomace (the skins and stalks left over after winemaking – 20–25 percent of the crop weight), mulch and vine prunings can all be used to improve organic matter too, returning nutrients to the soil ecosystem in a circular system rather than exporting them. At Lyrarakis, in Crete, for example, which has been an organic producer since 2020, they have stopped ploughing between vine rows and use only fertilizer made from pomace and organic waste from their kitchen and nearby oil mills.

Increasing soil's organic content in this way does more than just improve its health. It also increases the soil's carbon content, thereby making a difference to levels of carbon dioxide in the atmosphere. Launched by France at the COP meeting in 2015 and supported by the International Organisation of Vine and Wine, the '4 Per 1000' initiative indicates that an annual 0.4 percent increase in carbon stocks in the first layer of soil would significantly reduce atmospheric CO_2.

Prunings are another useful source of organic material. Traditionally, they were burned in the field, but instead they can be chipped and spread on soil, becoming part of the short carbon cycle as they decompose and return nutrients. Or else they can be converted into biochar, the form of charcoal produced by heating material such as prunings at high temperatures: this is then spread on soil or used as biomass fuel in place of fossil fuels. Biochar increases soil pH, moisture content and nutrient retention. It can also help rid soils of heavy metals and pollutants. At Tablas Creek, a makeshift kiln for biochar from prunings and tree cuttings sits next to the extensive compost area.

Dealing with pests

Environmental and climatic weather conditions affect susceptibility to pests and disease, as well as their spread. The vine is particularly susceptible to fungal diseases such as mildew, but other pests and diseases include bacterial diseases, viruses, weeds, insects, birds and mammals. Booker Wines in California suffers from gophers, burrowing rodents that damage young vines. At Jackson Estates' La Crema vineyard in Sonoma, California, wild pigs are a problem. Bears and wild boar venture into the vineyards at Magoutes, in a remote part of northwest Greece. And at Iona, in Elgin, South Africa, owner Andrew Gunn reports that they have to contend with a troop of around 20 baboons that lives nearby and are partial to grapes: Iona use farm worker patrols and firecrackers to discourage them. Kangaroos in Australia have been known to damage irrigation pipes to get to access to water when thirsty.

Pesticides vary from synthetically produced chemical formulations to naturally occurring substances such as sulphur, which has been used in agriculture for centuries. But whatever their nature, all pesticides are potent chemicals which risk killing beneficial life at the same time as the target organism. Pesticides reduce soil microorganisms' ability to transform unassimilable nutrients in soil into forms which can be absorbed by plants. This leads to an increased need for fertilizer, which in turn negatively impacts soil life – creating a vicious circle of diminishing returns. Pesticides can also leave chemical residues on plants, in crops and in soils, which are harmful to the land ecosystem and potentially to human health, and can get into water systems. Whatever the chosen method of farming, reducing pesticide use is an important goal.

Pesticides and their use are regulated and monitored by government authorities, which set limits for permitted residue – known as maximum residue limit (MRL). Growers are required to keep records of their use, and the handling of pesticides must be carried out wearing appropriate protective equipment by trained personnel.

Synthetic pesticides tend to have a greater negative impact on the environment. Contact pesticides remain on the outside of the plant and are only effective when in contact with the pest. This means that rain after spraying a contact pesticide washes it away, negating its effect and requiring another round of spraying if disease pressure persists. This is true of one of the most widely used fungicides, 'Bordeaux mix', a mixture of copper sulphate and quicklime/hydrated lime and water. Thus, for example, many English vineyards were hit by mildew in the wet summer of 2024. 'These wet summers make it very challenging,' says Simon Roberts, winemaker at Ridgeview, in Sussex. 'It's difficult spraying [fungicides] every two weeks when the ground is so wet.'

For example, Pyrimethanil is a contact fungicide used to combat powdery mildew (though it can have some systemic effects). By contrast, systemic pesticides, such as the herbicide Glyphosate (commercially known as RoundUp) or the fungicide Boscalid, are synthetic chemicals that penetrate and move within the plant. These can reach parts of the plant, like buds, that are difficult to treat: they generally remain effective for longer.

Vines' susceptibility to fungal diseases is the reason fungicides account for over 90 percent of all vineyard pesticide applications, and why the use of fungicides in viticulture is higher than for other agricultural crops. Powdery mildew, downy mildew and botrytis each have a characteristic life cycle, with resistant dormant spores that can survive for long periods over winter, within buds, on dead wood and on the ground: these spores become active when weather conditions become optimal for germination. Conditions for the development of each disease vary and susceptibility to each disease also varies according to the grape variety in question, but humidity, increased vegetative growth and a dense vine canopy all increase the risk of fungal disease.

Powdery mildew is the least dependent on weather conditions and is widespread around the world. It affects most varieties of *Vitis vinifera*, with Chardonnay and Cabernet Sauvignon particularly susceptible. Downy mildew is more common in regions with warm, wet springs such as Bordeaux and Australia's Hunter Valley. And botrytis occurs where there is high humidity and rainfall towards harvest time. Growers in humid climates use the most fungicides.

Bordeaux mixture was first used in Bordeaux in the late 19th century to combat fungal diseases. It remains widely used around the world, in particular for organic and biodynamic viticulture, but it does have limitations in terms of its effectiveness and environmental impact. It washes off in rain and requires repeated applications, each of which increases the number of tractor passes through the vineyard and the grower's carbon footprint. Also, the build-up of copper in the soil over time can lead to soil toxicity, with a negative impact on soil life and biodiversity.

Growers can reduce pesticide spraying by recognizing the symptoms of the main diseases, knowing their life cycles and the weather conditions that encourage them. For this, they need to monitor the vines for symptoms on leaves and shoots, then time the spraying for when the target organism is most susceptible, using the correct dose, when weather conditions are optimal. They can also use protective or preventative fungicides, applied before infection occurs, to block the germination of fungal spores. Rapid spraying in anticipation of infection as a preventative treatment is more effective than spraying after infection has set in.

Once a fungal infection has set in, only eradicant or curative fungicides can control the disease. Curative fungicides are mostly systemic, moving within the vine and reaching all parts of the plant. But fungal strains resistant to commonly used fungicides have now evolved. Chemical sprays should be applied sparingly and different products rotated to help prevent further pesticide resistance developing. Growers also need to mind their neighbours: windy conditions during spraying can cause drift. In one case, in 2014, 23 children at a junior school in Bordeaux's Blaye zone experienced nausea and headaches after a neighbouring vineyard sprayed fungicides on to its vines to treat downy and powdery mildew: some were hospitalized.

There are some natural alternatives to chemical sprays that include botanicals, products of mineral origin such as clays and sulphur, plant oils, lecithin and weed extracts. At Tablas Creek, the team combines sulphur spray with compost tea as a mildew deterrent in summer: it works to prevent mildew by outcompeting spores, and cuts the use of sulphur spray by 75 percent. Meanwhile, at Gundlach Bundschu in Sonoma, California, they use spore traps in the vineyard, constantly monitoring passively for botrytis and mildew spores. 'It's powerful from an economic perspective,' says Joe Uhr, director of winemaking.

Biological control, or biocontrol, is a means of controlling pests and diseases using living organisms or substances produced by organisms. For instance, pheromone traps are widely used in vineyards to attract European Grapevine Moths and prevent them from reproducing: male moths are attracted by a pheromone that they are fooled into thinking is produced by a female moth – and then trapped. Vine weevils can be controlled with biological pest control products such as Seeka, which releases tiny nematode worms that feed on the weevils' larvae. Johan Reyneke, in Stellenbosch, South Africa, uses the fungal microorganism T39 to make his vines more resilient to downy mildew. T39, a specific strain of the fungus *Trichoderma harzianum*, works in several ways. Firstly, it competes with pathogens such as downy mildew and botrytis for space and nutrients, thereby limiting their development. At the same time, it produces compounds that inhibit their growth. And, lastly, it triggers vines' natural defence mechanisms, making them less susceptible to disease. Though Reyneke reports that conventional producers in the area say it hasn't worked for them.

At Krontiras, in Luján de Cuyo, Argentina, winemaker Maricruz Antolin explains her long struggle with leafcutter ants, a headache for many Mendoza growers. The ants eat a fungus which, incredibly, they cultivate by feeding it with leaves and shoots. Growers say the ants can strip a whole row of vines overnight. Some growers, such as Catena Zapata, also in Argentina, try to dig out their nests, but these can be huge, several metres across and up to seven metres deep. Krontiras and other local biodynamic producers have been studying the ants for three years. 'The ants actually prefer native plants and flower petals to vines,' she says. 'There's more damage when you have less biodiversity.' Still, while she has grown more native plants and flowers around her vines, this still has not definitively removed the ant problem. She says that there is a predator spider that eats the ants, though it is still not clear whether it's possible to procure them commercially. 'This is the challenge of biodynamics,' says Antolin. 'You're looking for solutions in the ecosystem.'

At Booker Wines, Hilary Graves has found organic products ineffective against leaf hoppers, the main insect pest in Paso Robles, in California's Central Coast zone. Her solution is to drop lacewings, a natural predator, by drone into the vineyards twice a year. As she points out, it's a lot cheaper than chemicals too: 'One pass with a chemical insecticide costs $6,000 plus 80 man-hours, plus I'm compacting the soil and using fuels. Drone drops cost $4,000 and take just four man-hours.'

The other kind of pest in the vineyard is weeds. They can compete with vines for nutrients and moisture, as well as create an environment that encourages other pests and diseases to develop. So, in many vineyards, the practice has long been to keep soil between rows and between the vines bare by applying herbicides. Many weedkillers are non-selective, affecting a wide range of plants. They can also remain in soils for a long time after they have been sprayed, affecting the soil's microflora and polluting water systems. As they are toxic to a wide range of plants, they affect food sources for insects, birds and animals that feed on specific plant varieties. The long-term use of herbicides modifies which weeds grow back, encouraging what are known as tough-stemmed colonizer plants, which are difficult to get rid of and build up resistance to the herbicide. For example, the organophosphorus compound Glyphosate, the world's most widely used herbicide in settings from agriculture to managing roadside verges, is heavily used in vineyards. Concerns around its impact on human health and biodiversity have led to it being increasingly restricted. Especially if inhaled, it can cause respiratory problems and oxidative stress, and it has been linked to chronic kidney disease, gut disease and is, according to the WHO, 'probably carcinogenic'. In animals it can cause lethargy, hypersalivation, vomiting and diarrhoea, and it is especially harmful to aquatic and marine life. And in vineyards, it is harmful to soil life and earthworms, which are vital to soil health. Regions that include Prosecco and the Patrimonio appellation in Corsica have already banned its use; several European countries have announced future bans or massive restrictions on the use of Glyphosate. France has banned its household use, but still permits its use in agriculture.

One alternative to using chemical herbicides is to till the soil to break it up, and to mechanically remove weeds. This was, for many years, the preferred method for organic and biodynamic practitioners, but is now being called into question. Tilling is more expensive because it is more labour-intensive. Yet while it leaves no chemical residues, it does disrupt the soil ecosystem, damaging the complex network of mycorrhizal fungi and releasing CO_2 into the atmosphere. Tilling also damages soil structure, leading to erosion. Added to which, bare soils get hotter under the sun and have reduced water-holding capacity: when raindrops hit bare soil, the force breaks up soil aggregates and forms surface crusts, which leads to more water being lost as run-off.

By contrast, cover crops have long been used to reduce herbicide use and prevent erosion on sloping sites. Permanent vegetal cover also reduces soil temperature and allows moisture to be retained. Increasingly, growers choose to allow some grass or cover crops between rows and between vines. Cover crops increase the soil's organic carbon content. At Booker Wines, 'a lot of effort goes into the cover crop to support the cash crop,' says Hilary Graves, who plants a 16-species blend that includes turnips and daikon. She says: 'It needs to be palatable to the sheep, who graze it in winter.' Some growers argue against cover crops on the grounds that they compete with vines for water and nutrients. However, few cover crop roots penetrate deep into the soil. When water is limited, these are the first to dry out and die, while vines' roots naturally grow deep. Science has shown that competition for water in established vineyards is a bit of a myth, although there may be an impact on young vines until their root systems are fully developed.

'Nobody argues with the benefits of cover cropping,' says Alexandra Everson, sustainability analyst at Jackson Family Wines. Major trials run jointly by Jackson in California and Oregon and the University of California, Davis, are reaching similar conclusions. A five-year study under way since 2021 across 24 hectares is looking at the impact of regenerative viticulture on soil health, and the cycling of carbon and nitrogen. A further block of vines is now in the eighth year of a similar trial, concentrating on soil carbon and sequestration. It suggests that a no-till approach plus compost has worked best for deeper soil in the longer term: 'Minimal soil disturbance is key,' says Everson.

Biodiversity in the vineyard

Before Johan Reyneke's vineyards in Stellenbosch were farmed, like much of the rest of the area, they were covered by the Western Cape's native *fynbos* scrubland. When that is cleared, plants and animals that were part of that ecosystem disappear too. Yet this is in the heart of one of the most biodiverse regions on the planet: the Cape Floral Kingdom, one of six recognized floral kingdoms in the world, is home to over 9,000 plant species – and to 95 percent of the region's wine growing.

'People should be as concerned about diversity loss as about climate change,' says Reyneke, a pioneer of biodynamic farming in South Africa. So he has kept around 20 hectares of *fynbos* and is rewilding sections to join them up as a corridor, so that native animals will be able to cross the whole property. Already, he says, bucks and porcupines are coming back, and birds of prey are doing well. Insect life is vital too: wasps and ladybirds eat mealy bugs, the insect responsible for spreading leaf roll virus in the Cape.

Sometimes organic and biodynamic methods have unexpected consequences that demand creative natural solutions. Using wood-chip mulch caused weevils that threatened the vines; first, Reyneke put chickens in the vineyard and that solved the weevil problem, but these were targeted by birds of prey. 'We needed a hardy chicken,' he says – so they used a wilderness corridor to encourage guinea fowl, of which there are now hundreds in his vineyards. They are also the main predator of the snout beetle, which attacks vine buds, leaves, shoots and grape bunches. 'You get stability through diversity; it's the opposite of monoculture,' says Reyneke. 'Does it cost money? Yes, but so do chemicals.'

One of the goals of sustainable grape growing is to create a better balance between the farmed and natural environments: South Africa offers a particularly fascinating example of this. Not only is the Western Cape astonishingly biodiverse, but wine estates tend to be big enough to be able to rewild sections in a pattern that would be difficult in many European wine regions, where patchworks of vineyards touch one another and there are no wildlife borders or corridors between them. 'When you're an organic farmer in Burgundy, you don't contribute to biodiversity – it's too great a luxury to say, I won't plant this last half hectare,' says Carolyn Martin of Creation Wines. 'Here, you have the whole range of creatures. You can farm without harming the wildlife around you. We're spoiled – everything can live here.'

Thus, at Hartenberg estate in Stellenbosch, they have created a 65-hectare wetland around their irrigation dam, helping attract over 85 bird species. Hartenberg's Andrea Els says the wetland helps cool the air in the vineyards. And for over 20 years, the estate has used only biological pest control: they have owl boxes as homes for 18 breeding pairs, and perches above the vine canopy for birds of prey in the vineyards, to help keep down rodents, moles and snakes.

At Delheim estate, like Hartenberg a designated WWF Conservation Champion, they have planted some *fynbos* species as cover crops between vines. Volunteers have been hacking out alien plants such as blue gum trees as part of a programme to rehabilitate 150 hectares of *fynbos*. The reward is heavy use by animals of these corridors, including sightings of the endangered Cape leopard – though owner Nora Sperling-Thiel reports that porcupines have a troubling taste for eating irrigation pipes.

These are all good examples of how vineyards can encourage high levels of biodiversity both above and below ground. Areas surrounding vineyards can encourage it too. But there are other benefits for the health of the vines themselves. Complex landscapes where vineyards are surrounded by semi-natural habitats are less susceptible to pests and diseases. This is sometimes known as 'functional biodiversity'. Hedges and trees also create shade and help to keep temperatures low. Having animals, trees and flowering plants encourages beneficial insects such as predators of pests. Tablas Creek has its own 'insectary', an area of plants chosen to attract insects such as lacewings and ladybirds – white sage, manzanita, coyote brush, lavender, yarrow. Likewise, at Kaiken, in Mendoza, Argentina, agronomist Nicole Monteleone points out 'insect hotels', wooden boxes at the ends of some vine rows to help shelter insect life.

More biodiversity also increases the amount of carbon sequestered. Greater use of trees in vineyards is one promising approach now being trialled by some winemakers. For example, one of the winners of the UK Wine Society's 2024 Climate and Nature Fund, Languedoc producer Château d'Anglès, has a programme to integrate tree planting with its regenerative farming practices. Their goal is to plant 950 trees over three years, so that ultimately each vine is no more than 50 metres from a tree. Other Fund winners included Château Monconseil-Gazin, in Bordeaux, which wants to plant nearly a kilometre of new hedgerows around their vineyards, and Iona Wines, in South Africa, who will plant 1,200 metres of hedgerows. The latter project, where 700 trees are already planted, will protect young vines in this biodynamic vineyard from spray drift from the neighbouring property's apple trees.

The majority of vineyards are farmed using conventional and Integrated Pest Management (IPM) methods. Europe accounts for 84 percent of the world's total organic-certified vineyards, with Spain, Italy and France together accounting for 75 percent. France, Italy and Spain have the highest proportion of organic vineyards, representing 22, 20 and 18 percent respectively of their total plantings.

Taking France as an example, official figures for 2023 show that of its 789,000 hectares of vines, 22 percent were farmed organically by 10,739 growers. And of the 59,000 registered growers in France, 23,542 practised IPM with HVE certification (*see* page 82–83). Organic viticulture is easier in dry, warm regions – Les Baux de Provence AOC has been 100 percent organic since 2023, while the Côtes de Provence AOC has declared an intent to be 100 percent organic or HVE3 by 2030. But even in Provence, organic viticulture represented just 24 percent of total Appellation d'Origine Protégé (AOP) vineyards, with another 38 percent certified as HVE. And in some prestigious regions, such as Bordeaux and the Loire, which have a more humid climate, it is more difficult to work organically. Bordeaux currently farms 25 percent organically. However, disease pressure in 2023 and 2024 resulted in significant crop losses, forcing some growers to renege on organic certification. For them, it was simply economically unviable and the risk of loss too high. There are not many growers like Alex Dale in Elgin, South Africa, who plans financially for a failed crop one out of every four years. Even before the difficult 2023 harvest, a report in February 2023 by Banque Populaire-Caisse d'Epargne indicated that the proportion of French vine growers intending to go organic had dropped from 11 percent in 2021 to six percent at that time. Sometimes we forget that growing grapes is a business: even a family-run estate needs to make a living.

Meanwhile, organic and biodynamic viticulture require specialist knowledge and a high level of attention to detail in the vineyard. The average age of vine growers in France, at 53, is slightly above that in other branches of farming. Over half are aged 50–60, with many having learned their skills from father to son. And for those who did study viticulture more formally, organic, biodynamic or regenerative viticulture were unlikely to have been at the forefront of the curriculum. Organic and biodynamic farming involve getting out among the vines and not only being able to identify the warning signs of disease but also being able to react fast to spray each plot at just the right moment. With the best will in the world, realistically, it takes time for a grower to work around each vineyard plot. And as discussed on page 73, there is also the whole question of the use of copper over extended periods of time, leading to soil toxicity and increased tractor passes, which increase carbon footprint and compact the soils.

So the choice of farming method is not clear-cut and the examples we have looked at demonstrate some of the dilemmas growers face. Farming methods impact the cost of production. In Provence, Stephen Cronk of Mirabeau estimates that it costs up to €1,000 per hectare extra for organic viticulture compared to conventional. For organic and biodynamic, while there may be some saving on the cost of agrochemicals, more man-hours and manual labour are generally required, yields are lower and the risks higher. With conventional and IPM methods, growers have the flexibility of using synthetic chemicals at times of high disease pressure and may adopt organic, biodynamic or regenerative practices to a lesser or greater degree, but these cannot be certified.

As winemaker Sally Evans of Château George 7 in Fronsac, Bordeaux, explains: ‘I considered HVE better because I could use these biocontrol products which have the original molecule from the plant and then it’s concentrated with tech to go back on the vine to reinforce the plant’s natural defences. In a year like 2024, I did nine treatments, while organic farmers did 23–25. So, if you do everything else “organic” or regenerative – no chemical weeding, minimal intervention in the vines, planting hedgerows for biodiversity, light bottles produced locally with a high percentage of recycled glass and so on – that goes beyond “organic” to the bigger picture. I prefer to do that than get the organic label.’

The cost of production also varies according to plantation density, the age and state of the vines, the trellis and training system, and hand or machine harvest. Growers have to weigh up all of these factors against the prices they can sell their grapes or wine for. And this is against a backdrop of declining global consumption, which is being felt in many growing regions where supply is exceeding demand – for example, in 2024, 8,000 hectares of vines were pulled out in Bordeaux.

[*continues on* page 88]

Five Approaches to Farming Vineyards

Conventional viticulture

Conventional viticulture uses a traditional 'calendar' approach, where a grower sprays a set number of treatments at given times over the season. It's a belt-and-braces approach. Growers can use any permitted pesticides, synthetic agrochemicals or natural chemicals, as well as mineral fertilizers. They thus have the greatest range and flexibility of pesticides to select from: herbicides to control weeds and synthetic fungicides to control downy and powdery mildew and botrytis.

The costs of conventional viticulture can be high due to the price of chemicals. In the north Italian Piemonte region, for example, the estimated annual cost for controlling downy mildew alone in all conventionally farmed vineyards reaches as high as €16 million (US $18.6 million) a year.

'It's a mindset from the universities and chemical companies, buying a certain product for a given problem,' says Sebastián Tramón, head of sustainability at Viña Emiliana, Chile's largest organic producer. But the costs may be offset by the economic rewards of higher yields and lower risk.

Integrated Pest Management

Integrated Pest Management (IPM) uses the same chemicals as conventional viticulture but takes a more considered approach, using them in a more targeted way and combining them with other non-chemical methods to reduce the impact on the environment and people. Growers base their chemical applications on an understanding of disease life cycles, monitoring the incidence and severity of disease, and using predictive models based on temperature and weather conditions rather than systematic spraying.

Cutting the use of synthetic chemicals in this way should translate to cost savings. Growers using IPM may also complement chemical herbicides and mineral fertilizers with practices such as mechanical weeding to reduce chemical dose. IPM is promoted in the EU Directive on Sustainable Use of Pesticides (Directive 2009/128/EC), and is also fundamental to France's

Haute Valeur Environnementale (HVE) agricultural certification, and the Terra Vitis certification, specific to viticulture. There are 3 levels of HVE certification with HVE3 the most advanced. In 2023, 40 percent of French vine growers were certified HVE.

Organic viticulture

Organic and biodynamic farming are more restrictive, banning the use of synthetic agrochemicals in favour of natural chemicals, tilling and biological control methods.

This means organic wine growers have limited options to combat fungal diseases – so prevention is paramount. The pesticides they are allowed to use are contact chemicals which wash off with rain and so may need to be re-applied at regular intervals. For example, a conventional vineyard dealing with medium levels of downy mildew may carry out 12 treatments over the season, whereas an organic vineyard may require 20 passages of spraying. This is why organic viticulture can have a higher carbon footprint than conventional agriculture, due to the increased number of tractor passes needed in the vineyard.

Organic viticulture promotes the use of ecological processes and cycles rather than chemical sprays. It seeks to maintain ecosystems and the fertility of soils in the long term, increase biodiversity and protect natural resources. 'For us, organic is looking differently at what we do every day,' says Sebastián Tramón. 'We have to get rid of the myth about organic production that it's not as good as conventional,' says his colleague, Los Robles winemaker Noelia Orts.

Organic growers use sulphur to prevent and control powdery mildew infection. Sulphur protects vine leaves from infection and kills the fungus upon contact. But sulphur is also toxic to beneficial mites and spiders, including natural antagonists of the disease. Growers can reduce the dose of sulphur by using it with other biocontrol products and alternative preventative treatments such as whey, milk, potassium bicarbonate and canola oil.

More controversially, organic growers use copper-based preparations to control downy mildew. Copper hydroxide and copper sulphate solutions have been used in European vineyards since the end of the 19th century, but copper's use over time leads to a build-up of copper in soils to toxic levels, potentially damaging to soil microbes, worms, birds, mammals – and water systems. Rodrigo Barria, agricultural manager at Viña Montes, explains how, at their vineyards in Chile's Colchagua Valley, they stick to organic and regenerative principles in many respects: they plant cover crops and do not till the soil; they use no nitrogen fertilizer; they use compost tea with potassium, made with winery waste water; and the soil on their main site boasts a respectable 2.5 to three percent organic material. But they are not certified. He says, 'I'd like to be organic, but I don't like copper – it kills so many bacteria.'

In conventional viticulture, copper has been replaced by alternative synthetic chemicals. However, for organic and biodynamic growers, it remains the only permitted chemical to control downy mildew. It is a contact pesticide requiring repeat spraying after rain. Its use in the EU is limited to four kilograms per hectare, averaged over a five-year period. Some cultural practices help prevent downy mildew, such as avoiding the accumulation of water, avoiding working soils when wet, removing leaves showing symptoms, and lifting shoots off the ground early to avoid contamination. But there is no effective organic spray alternative at present. Research into biological pesticides using *Yucca schidigera* and *Salvia officinalis* extracts, as well as *Trichoderma harzianum* and algae-based sprays, remains ongoing. For botrytis, organic growers have few options other than cultural practices to prevent infection.

It adds up to a real challenge for organic growers in years with difficult weather. Alex Dale of Radford Dale in Elgin, South Africa, says that in their financial planning, 'we've built in a failed crop every four years – we have to, otherwise we're not financially sustainable'.

Different certification bodies exist between regions, with differing standards and approved treatment products. Some growers say that they practise organic viticulture but are not certified due to the cost of registration and/or a desire to retain the flexibility to use synthetic chemicals when absolutely necessary. In Maule, Chile, Derek Mossman Knapp's Garage Wine Co is not organic certified because, he says, it would be too expensive to do the paperwork for so many small sites of the kind he farms. 'Organic is a racket – the certification is a business,' says Knapp. 'I'd have to hire someone to do the paperwork and we're a five- to six-person company. But is it better if big companies go organic? Absolutely.' Yet without certification, such wines offer no guarantee to consumers.

Biodynamic farming

Biodynamic farming was developed in the 1920s by Austrian social reformer Rudolf Steiner. It is similar to organic in that no synthetic pesticides or fertilizers are used. But additionally, it is a holistic approach that considers the vineyard as a self-contained ecosystem, with minimal inputs from outside the farm. Treatments are applied according to an astronomical calendar based on phases of the moon, energy from the sun and moon, and the movement of other celestial bodies. Some winemakers reject the quasi-mysticism around Steiner's rules. Laurens Hartman at Karanika, northern Greece, essentially farms biodynamically but isn't certified: 'I follow nature, I don't follow silly rules made up 100 years ago in a different climate and a different history.'

South African biodynamic grower Johan Reyneke explains how he started: 'One day I said: "I'm done using poison." It was a disaster – within six months we had every pest and weed.' But gradually he taught himself biodynamic methods and his soil and vines began to thrive. As he summarizes his philosophy, 'nature works but no one gets up to farm it every morning'.

In biodynamics, growers promote soil life and vineyard health by having a wide diversity of plants and insects. 'The soil is the starting point for me,' says Alvaro Espinoza, Chile's pioneering biodynamic producer. Growers like him aim to continuously increase the soil humus layer, which builds up the water-holding capacity and fertility of the soil. Herbicides are not permitted; growers generally till to control weeds. Many such growers integrate animals into the farming system, supporting soil regeneration and providing compost. Biodynamic growers use preparations made from herbs, minerals and cow manure to stimulate soil regeneration and organism growth, applying them according to the lunar cycle. But biodynamic growers face the same challenges controlling fungal diseases using contact chemicals as organic producers, although with lower permitted levels of copper use. Wines made with biodynamic grapes are certified by Demeter in 45 countries.

Many believe that organic and biodynamic viticulture produce consistent but generally lower crop yields. The vines generally have less vigour and vegetation, and produce fewer bunches, smaller berries, thicker skins and less juice. 'The biggest challenge is that plants look leaner with biodynamics,' says Rosie Gunn at Iona, in Elgin, South Africa. 'The vines will never have the vigour of non-biodynamically farmed ones. But a pig running around in a farm is never going to look like a pig raised with feed.' A study conducted between 2008 and 2014 in McLaren Vale, South Australia, by the University of Adelaide, showed yield drops of 21 percent for organics and 30 percent for a simplified form of biodynamics, compared to high-input conventional farming. Conventional European growers who go organic/biodynamic also report that yields typically drop by 20 percent. However, research conducted at Geisenheim University, in Germany, indicated that while yields of organic and biodynamic vineyards were lower initially, in hot and dry years organic and biodynamic vineyards produced higher yields and better-quality grapes than conventionally farmed vines. In other words, the vines were more resilient and more consistent between years.

Regenerative farming

Regenerative farming is a term coined in the 1980s by the Rodale Institute, US, and has attracted much attention more recently. Regenerative agriculture implements a no-till approach, using permanent ground cover and sequestering carbon in the soil. Soil life is encouraged below ground as natural fungal hyphae networks are left undisturbed to build over time. Weeds are controlled through the planting of cover crops which are grazed by sheep and subsequently cut or rolled down. Plant matter and mulches reduce the amount of water lost through direct evaporation from the soil, as well as protecting against the heating effect of the sun. Trees and hedges planted in and around vineyards encourage soil life at deeper levels below ground and create habitat and shade for life above ground.

At Château Cheval Blanc in Bordeaux, which farms regeneratively, specific cover crops such as clover, mustard, radish, carnations, flax and wild rye are sown between vine rows after harvest. As these grow, over winter and spring, they prevent weeds and soil erosion and increase biodiversity above and below ground, providing habitat. In summer, the cover crops are mechanically flattened rather than cut, forming a blanket over the soil which continues to prevent weeds from growing, while also protecting soils from drying out. This helps to reduce soil temperature during summer heat and encourages rain infiltration.

The cover crop plants naturally die off in the autumn and decompose, returning nutrients to the soil. Fruit trees planted within rows of vines provide shade and encourage biodiversity, with birds, bees and other insects above ground and soil organisms below among the roots. Farming in this way requires manual work, as tractor use is restricted.

Regenerative agriculture shares similar principles to organic and biodynamic farming in encouraging biodiversity; many vineyard practitioners have adopted it as the next step on, having farmed organically or biodynamically for years. But a no-till approach can be applied to any farming method. Plus, regenerative farming doesn't necessarily prohibit the use of chemicals in the way that organic or biodynamic agriculture do.

There remain problems of definition – and indeed there is no internationally accepted standard for what exactly counts as 'regenerative agriculture'. The DFA of California, a non-profit organization supporting fruit and other agricultural producers, announced the first official definition of what constitutes regenerative agriculture in January 2025. Australian authorities are currently trying to agree a definition. And there are now four different certification schemes for regenerative agriculture: ROC (Regenerative Organic Certified, the biggest internationally), Regenified, the Regenerative Viticultural Alliance, and A Greener World.

One of the biggest differences between them is whether producers need to be certified organic before getting regenerative certification. It makes a big difference practically and economically for farmers, since it takes a three-year conversion period to become certified organic, during which yields are generally lower than under a conventional regime. Neither the Regenified nor the Regenerative Viticultural Alliance standards require producers to be certified organic, though as Josep Maria Ribas of Familia Torres points out, the latter scheme's auditing process does effectively require certified growers to be organic too – 'regenerative without organic doesn't make sense'.

Stephen Cronk, co-owner of Mirabeau in Provence (which operates France's only regenerative-certified vineyard) and a trustee of the Regenerative Viticulture Foundation, takes a slightly different view. 'I've changed my view on being organic first,' he says. 'You should be able to go regenerative without going organic – we need to meet farmers where they're at. Otherwise there's a danger that regenerative viticulture will be a niche within a niche.' Thus, under some regenerative certifications – though not ROC – farmers can use chemicals so long as they have a plan to wean themselves off them.

Adapting to climate change in the vineyard

We saw in Chapter 1 how climate change is threatening vineyards with warming temperatures, increased heat and water stress, and extreme weather events. Sustainable viticulture helps to build resilience to the effects of climate change, for example by increasing the soil's water-holding capacity, reducing pesticide use, and encouraging the slow natural release of soil nutrients.

Using water more wisely

Derek Mossman Knapp's winemaking in Maule, south-central Chile, is unusual in several respects. Mossman is from Toronto. He first fell in love with Chile when he went there skiing, as a student in the late 1980s. He moved to Chile after the return of democracy, in 1994, married a Chilean woman, and made his first Garage Wine vintage in 2001. His passion is old vines, especially Maule stalwarts Cariñena (Carignan), Monastrell (Mourvèdre), Garnacha and local variety País. He leases almost-forgotten vineyards of one to two hectares of such vines; court documents show that one plot of ungrafted Cariñena was planted in 1910. All are only accessible by 4x4; some are located down narrow forest tracks – the vineyard, a patch of paler green, almost surrounded by trees.

But in Chile, perhaps the most unusual thing about these vines is that they are dry-farmed, without irrigation. Knapp manages this because these are small plots where old vines have deep roots. 'Old vines give crucial insights into sustainability,' says Knapp. 'The old plant is actually more resilient – it has a memory of less pampered times.' In his view: 'You could dry-farm in the Maipo, but not on the current scale... You could plant using keyline [vines planted in sinuous rows], working with the contours more for better absorption of water.' But, he concedes, 'you'd never produce enough to make supermarket wine.'

The vast majority of vines in Chile, Argentina, and parts of Australia and South Africa are irrigated. In Argentina's arid Mendoza province, traditional systems of *turnos* – irrigation ditches – have for centuries channelled rivers of snow meltwater from the Andes to farmers' fields. At 1,600 metres up in the Uco Valley, the vegetation around Salentein's Las Sequoias vineyard is lush, with a magical wooded section nearby, which is home to Californian sequoias, poplars and quince and, close to a wide, meltwater-fed stream, cherry trees among the grass. But, says Salentein's agronomist, Lukas Mateu, the trees were planted by his grandfather in the mid-20th century. 'Otherwise, there would be nothing here,' he says, 'just bushes and low trees. Without water we can't do anything.'

Yet despite traditionally banning irrigation, many European wine appellations are increasingly requesting exemptions or changes in legislation to permit irrigation: 40 percent of Spanish vineyards are now irrigated. They face competition from other farmers: demand for water for other crops will increase as we see longer dry periods and lower water reserves.

Climate change makes it even more important for growers to use water efficiently. Using irrigation greatly increases water consumption. 'Every pump we have has flow meters on it,' says Joe Uhr, director of winemaking at Gundlach Bundschu in Sonoma, certified both organic and for regenerative agriculture. 'It's been a long time coming in the wine industry. County and state regulators turned a blind eye for the last 40 years. But water is becoming more and more of an issue in California. The Central Valley is now sinking by a measurable amount every year because of pumping by agriculture.'

Overhead sprinkler systems, a common sight in Australian vineyards in the 1990s, are the most wasteful irrigation method (they also promote fungal disease). Drip irrigation, which is now the norm in recent plantings in Chile and Argentina, was first developed in the 1990s. It uses water more efficiently, though it requires infrastructure, with the installation of pipes and hoses along each row of vines.

Nevertheless, even drip irrigation changes vines' behaviour. Irrigated vines' roots develop where the system delivers water: they do not search for water any further than the dripper. This is generally in the upper layers of soil, which is the first zone to dry out in drought, and the zone where weeds are most likely to compete. Irrigated vineyards become dependent on this supply of water: if reserves drop and irrigation becomes limited, they are less resilient to drought. So at Viña Montes, in Chile's Colchagua Valley, head of impact Josefina Astaburuaga says: 'We are trying to build up the plants' resilience, to get them to manage with less.' Montes have also experimented with lowering the vines' canopy height from 1.2 metres to 80cm, which led to a 13 percent reduction in water consumption. Dimitris Diamantis, owner of Magoutes, in northwest Greece, goes further: he believes that 'if you irrigate the vines, they become addicted to water'. His Xinomavro vines are dry-farmed. Irrigation can also lead to salt build-up in soils, with an impact on soil life. In addition, it pushes up energy consumption and associated emissions: that water needs to be pumped.

New technologies have made drip irrigation more efficient and are helping transform growers' understanding of their vines' water needs. Precision watering develops more resilient vines, as the root system is encouraged to spread. At Warwick Estate in Stellenbosch, South Africa, they use data gathered by drone to monitor plants' vigour and water stress. They also use neutron probes to measure the depth of the water level, allowing more precise irrigation. Similarly, at Jackson Family Wines in California, Craig McAllister, head winemaker at the La Crema vineyard, contrasts the old approach – 'it used to be, it's Monday, let's irrigate' – with the soil moisture probes at surface and root level that they use today to guide irrigation.

Meanwhile, at Symington estates in the Douro Valley, in Portugal, around 40 percent of their planted area is irrigated. Research and development manager Fernando Alves explains that their version of drip irrigation can reduce the amount of water used by up to 40 percent compared to what a conventional irrigation system would use. A new system of sensors on vines gathers data in real time, showing which places need water and how much. Symington now also give more thought to locations for planting in terms of how much water the plants will need, as well as choosing vine rootstocks better adapted to dealing with heat and water stress.

Another method is direct root zone irrigation, where water is delivered via a 60cm tube inserted in the soil, encouraging vine roots to descend. It also prevents water loss from evaporation and competition from weeds. This can reduce water use by up to 60 percent – at Ventisquero, in Chile's Maipo Valley, they think direct irrigation like this reduces their water consumption by 80 percent. Mike Ratcliffe, owner of Vilafonté wines in South Africa, says: 'Subterranean irrigation is becoming a really cost-effective tool, not only in reducing dependence on scarce water supplies, but also providing more effective vineyard management – and the increased cost in implementation is no more than 20 percent, with a 50 percent-plus reduction in water usage.'

Nevertheless, the long-term future of viticulture in some places that are heavily dependent on irrigation must now be in doubt. Already some Chilean growers are pulling up vines following 15 years of drought. 'We have 130,000 hectares in Chile, and a great number of those will be pulled out,' Sebastian Labbé, winemaker at Viña Santa Rita, told industry publication *The Drinks Business* in 2023. 'Many of them have not been farmed this year, and they only haven't been pulled out because that requires an investment.'

Sustainable viticulture maximizes rain capture and infiltration into soil, while trying to prevent run-off and loss through evaporation. Gerardo Leal, vineyard manager at Viña Santa Rita, talks of 'designing water-retention landscapes' in the Maipo Valley; choosing sites that don't drain too fast and using terraces or even roads to stop run-off.

To reduce overall water consumption, producers can also use recycled and treated winery water, as at Gundlach Bundschu. At Jackson Family Wines in California, they now keep large, open-top, stainless-streel tanks outside when not in use, collecting some of the approximately 1.9 million litres gathered from rainfall across the company. This is used in irrigation and in refrigeration in outdoor cooling towers.

Indeed, cutting water inputs is one of the arguments in favour of techniques that focus more on the health of a vineyard's soil, especially biodynamic and regenerative viticulture. A one percent increase in organic matter can increase water-holding capacity by 3.7 percent. Put another way, a US study showed that every extra percentage point of organic matter in an acre (0.4 hectares) allows the soil to retain another 75,700 litres of water, because of its effect on soil structure. Similarly, comparative studies by some California wineries on regenerative versus conventional agriculture show that cover crops between vine rows can significantly reduce both the soil surface temperature and the temperature of the leaf canopy, thereby reducing heat stress and the plants' need for water. While not yet widely adopted, biocontrol methods that encourage vines to use water more efficiently while reducing water and thermic stress are also available.

As we have seen, although there are many aspects to growing wine grapes, there are guiding principles that can be applied whatever the context. Sustainable viticulture minimizes the use of all pesticides – in particular synthetically produced agrochemicals – encourages biodiversity within and around vineyards, increases carbon levels in soils, works towards natural resilience to pests, disease and climate events, and uses water wisely.

Renewing vineyards:
what happens when farmers plant new vines?

Establishing a vineyard is a significant, long-term investment. The land must be cleared and prepared before planting, and it takes two to three years after planting before the first grapes are harvested. During that time, the vineyard work continues: vines must be pruned each year and trellis systems installed. The cost of the vines and infrastructure – trellis posts, wires and irrigation systems – varies substantially by region, according to density, and the trellis and training systems in question, but is of the order of $50,000+ per hectare, not including the cost of the land.

In addition to the financial cost, there is an environmental cost in establishing a vineyard. Preparation of the soil, clearing land, ploughing and fertilization, installation of trellis and irrigation systems, and the manufacture of any materials used: all have an environmental impact. If a grower has cleared virgin land and disrupted natural ecosystems – as, for example, with recent plantings in Mendoza's Uco Valley, in Argentina – that has a higher impact than converting land previously used for other crops. However, vines are long lived, and the longer a vineyard is productive, the higher the return on both the original financial investment and the environmental cost.

Vineyards are replanted periodically; as crop yields decline with vine age or disease a point is reached at which they are no longer economically viable. To design and manage vineyards that will be productive for many years, with the uncertainty of climate change, demands that a grower builds resilience into the new vineyard. So producers need to consider the materials used to build trellis and irrigation systems, to reduce the environmental impact of their manufacture and installation, and provide maximum longevity.

Growers also need to make tricky decisions about which grape varieties will be most suited to the climate in a given location in 10 or 20 years' time. Researchers around the world are researching grape varieties, rootstocks and heat-and-water stress mechanisms. Bosman Wines in Stellenbosch also operates a major nursery that produces a third of all South Africa's grafted vines: research on heat- and disease-resistant vines is part of their work. Owner, Petrus Bosman, explains that experiments with different clones of vines are one strand of this. For example, Chenin 258 has looser bunches that offer better protection against rot: mildew pressure here is relatively high. But he has also planted test rows of more heat-resistant varieties such as Tempranillo, as well as hybrids like Sirius.

The latter are a significant area of such research in Europe. So-called PIWI (*PilzWiderstandsfähig* – a German term) grape varieties are crossings of *Vitis vinifera* with American or Asian species such as *Vitis berlandieri*, *Vitis riparia* and *Vitis rupestris*. PIWI varieties are bred for their resistance to fungal diseases, thereby reducing the amount of spraying required. In 2024, UK supermarket Tesco launched the first mass-market wine made from one of these hybrids, Floreal, created by crossing Villaris (a PIWI variety derived from Sirius and Villard Blanc) with a descendant of *Vitis rotundifolia* (or 'Muscadine', native to the south-central US). The grapes in Tesco's Floreal wine are grown in the Languedoc and the Loire. The new variety is highly resistant to fungal diseases such as oidium and mildew: it reportedly requires up to 90 percent fewer vine treatments (and as a result, less use of tractors). Critical reception of the wine has, however, been mixed.

One solution to climate change's warming effects is to plant vineyards at higher altitudes or cooler latitudes. As a rule, temperature drops approximately -0.6°C for every 100-metre increase in altitude.

Higher average temperatures have made higher sites more viable: in Burgundy, for example, vines in the higher, cooler Hautes Côtes de Bourgogne now ripen grapes more consistently than in the past. In South Africa, rising temperatures have led growers to seek out cooler areas such as Elgin, just a few kilometres from the Atlantic Ocean. Radford Dale worked only in Stellenbosch for their first 20 years but then bought a biodynamic vineyard in Elgin. 'We future-proofed the business by investing here,' says owner Alex Dale. 'There's a month difference in picking times.'

Wine growers in Chile and Argentina especially have pioneered ever-more-extreme locations. In Chile, both Montes and Santa Rita have planted grapes on the island of Chiloé, in the south of the country, which sits on the 42nd parallel south, at a similar latitude to Tasmania and Marlborough, New Zealand. In Argentina, the Avinea Group has launched Bodega Otronia with vineyards planted in Patagonia, at a latitude of 45 degrees south. It's the southernmost winery in the world: the temperature there can drop to -20°C in winter. The vineyards are surrounded by lines of trees, planted as windbreaks: there is an average year-round wind speed of 30 kilometres per hour (km/h), sometimes topping 100km/h.

But while such places can make interesting wines, they are unlikely to alter the shape of the wine industry in those countries much. Noelia Orts, winemaker at Viña Emiliana's biodynamic Los Robles estate, says: 'I don't think the solution is just to move to the south. We have a vineyard in [the southern region of] Bío Bío, where we get 1,000ml of rain a year, but we need solutions in other places.' Moving vineyards in this way also has downsides in addition to its impact on ecosystems. There can be a social and economic cost from the impact of lost jobs – and what happens to the abandoned vineyards?

Sustainable viticulture uses resources efficiently, adapting existing vineyards and mitigating climate change with techniques like field grafting alternative scion varieties, which can better resist high temperatures and drought, onto existing vines, or using alternative trellis systems and canopy management to protect grapes. One approach is massal selection, where viticulturists regenerate a vineyard by taking cuttings from old vines rather than using a single clone, thereby preserving the genetic material of a biotype adapted to a micro-terroir and improving the vineyard's resilience. As Bénédicte Gallucci, viticulturist at organic southern Rhône producer Chêne Bleu, explains: 'We take cuttings from the best vines, rather than bringing in commercial clones. We aim to select the low-yielding vines to give us the quality that comes from concentrated fruit.'

Conclusion

Growing grapes is the fundamental base of all winemaking. The sensitivity of *Vitis vinifera* to climate conditions dictates which varieties can be grown where and makes the vine increasingly susceptible to the uncertainties of climate change, which is now affecting grape quality and yields. Viticulture is the canary in the coal mine: what we see happening in vineyards will affect other forms of agriculture too. The adaptations and mitigations that are being developed in response to a wide range of environmental implications are applicable outside the arena of wine.

The next stage in the process – the transformation of those grapes into wine – presents just as complex a situation, with a completely different set of challenges. That is the subject of the next chapter.

General lessons for sustainable agriculture

- **Focus on soil health**, especially on increasing the soil's carbon content, by applying the principles of regenerative agriculture.
- **Minimize chemical use**, in particular synthetic agrochemicals such as some fertilizers and pesticides.
- **Use biological control methods** where possible, to control pests and stimulate vines' natural defences.
- **Use water wisely**. If you have to irrigate, do so using technology that limits water use to a minimum.

› **Use cover crops and mulches** to reduce soil water evaporation and reduce soil temperature.

› **Minimize energy use** – both electricity and diesel – and use renewable energy where possible.

› **Encourage biodiversity** and develop natural ecosystems in and around your vineyards. It will benefit the vines as well as nature.

› **Adopt a circular approach**: return prunings, compost and mulches (carbon and nutrients) to the ecosystem.

Chapter **THREE**

MAKING GRAPES INTO WINE

Standing in her winery in Luján de Cuyo, Argentina, Maricruz Antolin, winemaker at Krontiras, is tasting samples straight from the tanks with a group of friends one evening. It is mid-March and the harvest is not yet complete, though she already has a row of small stainless-steel tanks full of fermenting wine. A diminutive but charismatic figure in ponytail, jeans and last year's Krontiras harvest T-shirt ('Keep calm and ferment with wild yeast') she moves from tank to tank – 'You guys have to taste this!' Her still-emerging wines are bold and different, including an orange Chardonnay and a no-sulphur Malbec: Antolin makes her wines from biodynamic grapes and uses few interventions: as her shirt proclaims, she uses only natural yeast and adds no sulphur dioxide (SO_2) as a preservative to most of her wines.

It is a rather different scene during the 2025 Southern Hemisphere harvest a few weeks earlier at Leeuwenkuil Family Vineyards in South Africa's Swartland region. Head winemaker, Pieter Carstens, stands between a row of huge, stainless-steel tanks, the smallest 50,000 hectolitres, the largest holding 350,000 hectolitres of must (grape juice). Down here, we are below ground: the winery is built into the earth to keep it cooler, and a vast array of solar panels on the roof supplies all its energy needs. This state-of-the-art facility opened in 2020 and can process 30 million litres of wine a year. Machine-harvested grapes picked the night before arrived at 10am this morning and are now in the nine huge presses: the winery can press 1,000 tonnes of grapes a day. Carstens is a towering Afrikaner in boots, shorts and logistics-company T-shirt: he takes a glass of Sauvignon Blanc juice from a tank, tastes and spits into a narrow drain running the length of the floor. He shrugs: 'We generally use commercial yeast.'

Whatever the contrasts between wineries, all winemakers face similar challenges in harnessing the microbiology that causes fermentation to make the best wine they can. While humans have been making wine for millennia, it wasn't until Louis Pasteur revealed the world of microbes in the 1850s that the transformation of grapes into wine was understood as a biological process rather than some kind of alchemy. Pasteur showed that living yeast microorganisms convert sugar into alcohol. It is essentially a biochemical process, like making bread, yoghurt or sauerkraut. During the alcoholic fermentation, yeast use sugar as a food source for energy to grow and multiply, converting it into ethanol while releasing carbon dioxide and heat. Pasteur also showed that other microorganisms cause spoilage which can turn wines sour and vinegary, develop off odours or make them become cloudy.

As the main ingredient in wine, grape quality has a direct impact on wine quality: you cannot make great wine from bad grapes (though, sadly, it is quite possible to make bad wine from good grapes). So a lot of effort goes into growing grapes. Alcohol, acidity, sugar and, in the case of red wines, phenolic compounds, and the balance between them, determine the balance of a wine, while aromatic compounds and precursors in grapes express themselves in wine flavour. As we saw in Chapter 2, grape sugar, acidity, aromatic flavour and phenolic compounds are all determined by growing conditions – and climate change is having an impact on those, principally because of what higher temperatures do to the ratio of sugar to acid, flavours and tannins in grapes. But whatever the state of the picked grapes, it is the role of the winemaker to maximize their potential in the winery. The winemaker's main tools are winery hygiene, temperature control, sulphur dioxide, acid adjustment, oxygen management and microbial inoculation.

Sustainable wine production is about maximizing wine quality using all resources efficiently (grapes, water and energy) while minimizing greenhouse gas emissions, waste and pollution – as well as creating a safe environment for winery staff and surrounding communities. Wineries use a lot of electricity for refrigeration and water for cleaning, particularly at harvest time. And they create waste: organic waste in the form of grape skins, pips and stalks (pomace) – as well as non-organic waste in the form of used filtration materials and packaging. Wineries generate greenhouse gas emissions mainly from their energy use, but they also directly emit carbon dioxide during the alcoholic fermentation.

The winemaking process

To start, it's helpful to understand the winemaking process. This can take anything from a few days to several years to complete. And while the basic process of alcoholic fermentation is the same whatever the wine, winemakers arrive at different finished wines in many ways.

To make still white and rosé wines, grape juice is fermented at a relatively low temperature – about 15–22°C. White wine is generally produced from white grapes and rosé wines from red (black) grapes. The grapes are normally pressed and chilled within hours of picking, either immediately or otherwise after crushing and a short maceration on skins. The colour of rosé wines depends on the amount of time the juice remains in contact with the grape skins, which give its pink colour. At this stage winemakers normally add SO_2 and chill unfermented juice from the press as it is pumped to a tank, to prevent oxidation and the onset of fermentation.

Freshly pressed grape juice, like fresh apple juice, is cloudy, thanks to the bits of pulp and skins in suspension. So it is usually clarified prior to fermentation, to avoid vegetal or earthy characters and to maximize the wine's fruit character. Like freshly pressed apple or orange juice in your kitchen fridge, solid matter settles to the bottom of the tank, leaving clear juice above a sediment. The separated sediment may subsequently go through a filter press to retrieve maximum juice. Other ways of clarifying include centrifugation, flotation and/or the use of pectolytic enzymes.

Many white and rosé wines are then fermented at 14–16°C in temperature-controlled tanks made of stainless steel or concrete. Fuller-bodied white wines tend to be fermented at slightly higher temperatures – 20°C – and may be fermented in oak casks or vats, or clay amphorae. How long fermentation takes depends on temperature, yeast strain and alcohol content, but it is usually done in a week or so.

For red wines, it's a little different. Red wines are produced from black grapes with the juice usually fermented in contact with the grape skins. Black grape skins and seeds contain phenolic compounds, including anthocyanins and tannins, which give red wine its colour, character and structure. So, unlike whites and rosés, with reds the skins and pips are left in the grape must – though in most cases the stems are removed first. Red wines are generally fermented at around 30°C. The alcoholic fermentation produces CO_2, which bubbles up through the must, lifting the grape skins to the surface of the fermenting mass to form a layer floating on top known as the 'cap'.

As it is the skins that contain the majority of the colour and tannins, winemakers need to use a variety of methods to ensure that these are extracted. In 'pumping over', fermenting must is drawn from the bottom of the tank using a pump and poured back in at the top of the tank over the cap. As the juice then slowly seeps down through the cap, it absorbs colour and tannins from the skins along the way – it's a process similar to the stove-top coffee pots where water bubbles up and back down through the coffee grounds. Alternatively, a winemaker can use *délestage*, otherwise known as 'rack and return', where the juice is pumped into another vessel, leaving the skins in the tank. Then they pump the fermenting must back on top of the mass of skins. *Pigeage* ('punching down'), involves plunging the cap down into the juice, either mechanically or by hand.

At the end of the alcoholic fermentation, red wine is separated from the skins into another tank. The skins are then pressed to produce 'press wine'. Press wine is usually more tannic and lower in acidity than the free-run wine. A winemaker may incorporate it with the free-run wine as part of the final blend, or keep it apart. Press waste consists of decomposed grape skins and pips – known as pomace or *marc* – and contains alcohol: in Europe the pomace is mainly sent for distillation into industrial alcohol, though it can be used for composting too.

Sweet and sparkling wines, wines made from dried grapes, and fortified wines all start with a similar process followed by additional steps. Sparkling wines generally undergo a second fermentation by which they get their bubbles, this may be in tank or bottle. Fortified wines are made by adding alcohol spirit.

Whatever the desired final wine, at the end of alcoholic fermentation, yeast cells sink to the bottom of the fermentation vessel, creating a slurry that is high in organic material and solids. This is known as the wine or yeast lees. As the yeast cells break down, they cede substances that give viscosity and weight, which increase the wine's mouthfeel and flavour complexity. This is why winemakers mature some wines on lees, stirring them up periodically, to maximize enzyme activity. But once the alcoholic fermentation is complete, winemakers often carry out an initial 'racking' to separate the wine from the 'heavy' lees. The young wine is pumped into another vessel and left on 'fine' lees, thereby reducing the risk of reduction and off odours. They usually add SO_2 to white and rosé wines to reduce the risk of spoilage during maturation and storage.

Most red and some white wines also undergo a further microbiological process before ageing known as malolactic fermentation. Malic acid, naturally present in grapes (also found in apples), has a tart, sour, green flavour. Malolactic fermentation converts this acid into softer 'lactic acid' and CO_2, thus reducing the wine's acidity and increasing its complexity, as well improving its microbial stability. This process is carried out by lactic acid bacteria (LAB). LAB are sensitive to temperature, alcohol and SO_2, so for wines where a winemaker wants to avoid malolactic and keep more acidity, it can be prevented by adding SO_2 after the alcoholic fermentation. Where a winemaker does want malolactic fermentation, LAB develop at an optimum temperature of 20°C – so because grapes are usually harvested towards the end of summer and the alcoholic fermentation finishes in autumn, when temperatures are low, completion of malolactic can take several months. During this time, the wine must be left un-sulphited, making it more susceptible to spoilage. Some winemakers heat wine or increase ambient temperature in the winery to encourage indigenous LAB development, while others inoculate with a selected strain of *Lactiplantibacillus* or *Oenococcus oeni*.

So while the alcoholic fermentation is the basis of the grapes' transformation into wine, other microbiological processes also affect the final wine's quality and stability. And the risk of spoilage from oxidation or microbial spoilage is present in grapes and wine at all stages in the process – which can significantly reduce its quality and value.

Maturing wine

Walking around the winery at Domaine Bousquet, in Argentina's Uco Valley, chief winemaker, Rodrigo Serrano, sketches the range of methods by which they mature wines after fermentation. There are around 200 tanks in this big winery, roughly half of them stainless steel and half concrete: while some of these are already being used for this year's harvest (which, in mid-March, is about halfway through), others are filled with maturing wine, which will rest where it is for up to a year. There is also a range of French oak barrels, 225-litre Bordeaux *barriques* and 400-litre *foudres*, and some larger concrete eggs. The eggs, like barrels, can be used for fermentation as well as ageing. Finally, Serrano proudly shows off his 400-litre Clayver ceramic Italian barrels, which allow micro-oxygenation without the wine getting any oak influence: 'We want to put the place in the bottle, not oak,' he says. However, at €8,000 apiece, they are pricey, even compared to French oak (it's €3,000 for a *foudre* the same size.) So Bousquet has just five of them to date.

Wines may be matured for anywhere from a few weeks to several years before bottling. Inert vessels made from concrete or stainless steel impart little flavour. But semi-porous materials such as Bousquet's oak barrels and concrete eggs – or clay amphorae – allow the exchange of tiny amounts of oxygen and other gases between the air outside and the wine in the container, which react with compounds in the wine to affect how it matures. In addition, oak imparts flavours such as vanilla, toast or roasted coffee, depending on the type of oak (French or American), the age of the barrels (the older the barrel, the less flavour it imparts), and how they were made (the barrels' interior is charred during their manufacture, to various degrees, hence they'll be described as having 'medium toast' and so on).

During maturation, wines may be decanted, filtered and clarified. Post-fermentation wines are cloudy, with solid matter in suspension. With time, heavier matter sediments to the bottom of the maturation vessel. Colloidal matter is also present: this is not visible to the eye but can cause haziness in bottled wine over time and unfortunately cannot be removed by filtration alone. To get rid of it, winemakers use fining agents that react with colloidal compounds to produce larger compounds which are heavy enough to settle, or are filtered out, leaving a bright, clear, stable wine. Fining agents traditionally include animal protein products such as albumin (egg whites), gelatine and isinglass (collagen obtained from the dried swim bladders of fish). Bentonite clay is often used for fining white wine and, increasingly, plant-based alternative products are available, such as plant-based gelatine and pea protein suitable for vegans.

A winemaker may also take action to remove tartrate crystals. Tartaric acid is the main acid naturally present in grapes. It reacts with potassium and calcium in wine to produce potassium bitartrate and calcium tartrate, which can form tartrate crystals in fermented wine. Consumers can confuse these with glass fragments at the bottom of the bottle, so wines are generally treated to remove them: this involves chilling the wine to a low temperature to encourage the crystals to form, then filtering them out.

Filtration in general is a subject of much debate among winemakers. Some believe it strips a wine of character: accordingly, natural and other low-intervention wines are often labelled as unfiltered. Others use only very loose filtration. Filter materials such as kieselguhr (a powdered sedimentary rock), sheet filters and membrane cartridges also create solid waste outputs from the winemaking process. Any sustainable winemaker needs to consider how to dispose of these.

Finally, wines are bottled, either at the winery on a permanent bottling line, or a on mobile line operated by a contractor, or transported in bulk and packed in another location. Packaging has major environmental impacts in its own right, which we will look at in the next chapter (Chapter 4).

Winemaking tools: battling microbes and oxygen

Rodrigo Serrano at Domaine Bousquet emphasizes the importance of hygiene in the winemaker's battle against harmful microbes. 'As a primary step, our team relies on careful observation to assess the specific cleaning requirements for each piece of equipment', he says. He continues: 'In instances where there is a significant build-up of adhered residue, we employ a specialized high-pressure cleaning system. Following this thorough cleaning, we perform a disinfection process to ensure complete hygiene. Conversely, if a tank has been used for wines that have undergone filtration and sterilization, our cleaning protocol involves a straightforward rinsing with water, followed by steam disinfection to guarantee microbiological safety.'

Winemakers are up against a range of potential problems – of which one of the biggest is microbial spoilage. This is simply because the natural environment around us is teeming with microscopic life – bacteria, yeasts and fungi – which are present in vineyards, on winery equipment and on grape skins. So, crushed grapes and juice are a melting pot of both good and bad microbes – though in terms of micro-organisms in the winery, we are really talking about yeasts and bacteria.

Each strain of yeast or bacteria has its own set of sensitivities and tolerances to the conditions that surround it. These determine whether it thrives and multiplies; simply survives, maybe in a dormant form; or dies. These tolerances are used by winemakers – and indeed in food manufacturing generally – to encourage the microbiological processes that they want to happen, prevent spoilage by the microbes they don't want, and to achieve stability. Factors that determine or limit microbial growth include alcohol, sugar concentration, temperature, aerobic/anaerobic conditions, SO_2 levels and the presence of other microorganisms. Lower acidity in grapes both affects the final wine balance and increases the potential for microbial spoilage: this is why, in hot regions, grape must is often artificially acidified. Acid adjustment is permitted, and is often standard procedure in places like California and Australia. The most common acid used is tartaric acid, but malic and citric acid are also possibilities.

Then, as fermentation proceeds and alcohol and CO_2 are produced, they alter the profile of microorganisms favoured. Microbial spoilage can show up in many unpleasant ways. Acetic bacteria can develop under aerobic conditions and produce vinegary aromas. *Brettanomyces* yeast and some lactic acid bacteria create horsey, barnyard aromas. And a variety of other bacteria can create off flavours too. Meanwhile, grapes and wine are also susceptible to oxidation if they are exposed to too much oxygen, even though CO_2 produced by fermentation protects against this to some extent. Oxidation can occur at any stage, making wines become dull and brown in colour and lose their fresh fruit aromas. So the type of spoilage risk will vary throughout the winemaking process.

This is why winery hygiene, like that at Domaine Bousquet, is vital. Wineries use water for cleaning at every stage of the winemaking process, but the peak water consumption time is around harvest, when harvesting machines, hoppers, tanks, barrels, presses, pumps and hoses must all be constantly washed down. Each piece of equipment has to be cleaned before and after every use. All structural areas, walls and floors must be kept clean. And all surfaces in direct contact with the wine must be rinsed with potable water to eliminate chemical residues. In fact, wineries consume anything between half a litre and 10 litres of water per litre of wine produced – and that doesn't include water used in the vineyard and in bottling. Most winery water is used at an ambient temperature, but bottling operations require hot water, at 60–90°C, or steam for sterilization of equipment and regeneration of filters.

Temperature control is also essential in the battle to control microorganisms: it accounts for the majority of winery energy use. The temperature of the environment in which microorganisms find themselves affects their ability to multiply, survive or die. This is why we leave bread dough in a warm place to rise – so that yeast can do its work. Conversely, we keep food in the fridge to inhibit bacteria that will make it go bad. Likewise, when a winemaker controls the temperature of grape juice, fermenting must and wine, they encourage, slow or prevent microbial development and associated biological processes.

So harvested grapes and pressed juice are chilled to prevent the development of spoilage organisms and to avoid untimely spontaneous fermentation. The alcoholic fermentation produces heat, often making refrigeration necessary to maintain a constant temperature. Fermentation temperature affects yeast metabolism, and the flavours produced. Most white wines are fermented at temperatures below 20°C to promote and retain fresh aromatic compounds. Red wines are often fermented warmer, to enable the dissolution of colour and tannins for wine structure. However, if temperatures become too high or too low, it can affect the yeast, resulting in a sluggish or stopped fermentation. But winemakers also use this phenomenon to advantage with rapid chilling to arrest fermentation when making off-dry or sweet wines. Finally, most wines are kept at a constant cool temperature during maturation.

Winemaking tools: the row over sulphur dioxide

A key tool used by most winemakers is sulphur dioxide (SO_2). However, sulphur has become controversial over the past decade: it is one of the principal targets of 'natural' winemakers. So why the controversy?

Sulphur dioxide is a widely used food preservative with anti-microbial, antioxidant and anti-oxidasic (inhibiting enzyme oxidation) properties. It is used in the production of juices, fresh fruit salads, dried fruits and certain meat products like salamis. In fact, yeasts naturally produce low levels of SO_2 during alcoholic fermentation – so no wine is actually free of sulphur – though not in concentrations great enough to give the wine much protection against spoilage. Thus, in winemaking, SO_2 is commonly used to prevent spoilage of juice before fermentation, as well as to protect the wine during maturation, storage and bottling: winemakers usually add it at several stages in the winemaking process. SO_2 prevents wine oxidation and restricts the growth of microorganisms.

For centuries, sulphur was burned inside rinsed oak casks. The SO_2 produced dissolved in residual water, sterilizing the barrel. Today, it is normally added to grape must or wine in the form of potassium metabisulphite or potassium bisulphite solution. Red wines contain polyphenols, which are natural antioxidants that protect wine from oxidation, so they generally need less SO_2. White wines, which are free of polyphenols, need slightly higher levels. Sweet wines containing residual sugar, which naturally binds with SO_2, are permitted to use more.

A small number of winemakers argue that SO_2 gets in the way of their grapes fully expressing their terroir. Maricruz Antolin refuses to use added sulphur, in line with her low-intervention winemaking philosophy. 'We believe this is the only way to get the true expression of our wines,' says Antolin. 'When we add sulphites at the beginning of fermentation, we lose aromas and flavours. To avoid SO_2 we work with very healthy grapes and try to keep the natural acidity. That's why we harvest very early.' British journalist and natural wines specialist Simon Woolf agrees: 'Sulphur mutes the wine… it gets in the way of fruit expression.' He is pragmatic about some sulphur sometimes being necessary, but adds: 'When the grower gets it right and the fruit is perfect and ripe, and the cellar is clean, you can get this extraordinary vibrancy and energy in the wine which is very difficult if you're adding sulphur.'

Elsewhere, some winemakers seek to reduce the amount of sulphur that they add to a minimum. At Bodegas Félix Callejo, their approach is low intervention, with only natural alcoholic and malolactic fermentations. However, says winemaker Noelia Callejo: 'At the beginning, we add a little metabisulfite to protect the grapes, prevent bacterial growth, and promote fermentation. But these doses are very low.'

Yet for some natural winemakers, SO_2 has become a near-obsessive focus: most natural wines use little or no added SO_2. While there is no legal definition of natural wine, the implication is that this approach is healthier. Making wine without added sulphur has become a prime goal of many natural winemakers, especially in France. But making wine without sulphur risks undesirable microorganisms producing off aromas and uncontrolled oxidation.

And despite their rhetoric on the evils of SO_2, levels of it in wine are comparatively low. There are legal maximum levels for total SO_2, which vary, but are far lower for wine than for other foods, such as dried fruit or cured meats. Wine contains on average 10 parts per million (ppm) of sulphur compared to up to 1,000 ppm for some dried fruit, especially apricots. It is true that SO_2 can cause an allergic reaction in some individuals – but it generally isn't dangerous unless breathed in at high levels. Conversely, recent studies suggest that low-intervention wines can have an increased concentration of allergenic substances such as biogenic amines due to the uncontrolled development of spoilage organisms. And there is no truth at all to the folk myth that sulphur in wine causes headaches: you get a headache from drinking too much alcohol (the tannins in red wines can make this worse).

There is no one substitute for SO_2 that offers its combined antioxidant, antiseptic and chemical properties. However, there are alternatives which can help to reduce the quantity of SO_2 that needs to be added. Most importantly, SO_2 use is reduced with good-quality grapes and good winery hygiene. Winemakers can use inert gases such as CO_2 and nitrogen to protect against oxidation, for instance to purge air from pipes and tanks prior to filling, or as a blanket over the wine surface in a tank of wine. They can use dry ice and CO_2 blankets to protect incoming grapes and juice. They can also use biocontrol agents to limit the development of unwanted microbes. These involve introducing positive microflora to compete with the population of spoilage organisms: for instance, *Metschnikowia pulcherrima* is a non-fermenting yeast that protects against oxidation as it consumes oxygen and controls the development of a wide range of undesirable microorganisms.

Winemaking tools: yeast, microbiology and oxygen

For Noelia Callejo, co-winemaker at Bodegas Félix Callejo, in Spain's Ribera del Duero, the choice of yeasts to make her wines is simple. 'All our yeasts are natural,' she says. 'Yeasts are naturally present on the skins of grapes – it's just that natural yeasts are less easy for winemakers to control than commercial yeast strains. For us, fermentation is spontaneous. It's slower, little by little, and at low temperature: it starts at 13°C and is never above 23,' says Callejo. She does not see herself as a natural winemaker. However, she says of natural yeasts: 'It's a way of respecting the landscape – it's important to us.'

But modern winemakers *do* have a choice: many use commercial dried yeast. The use of selected yeast and bacteria strains are the root of some of the biggest disagreements in modern winemaking, between those producers who favour use of commercially multiplied yeast and bacteria preparations and those championing low intervention.

For wild yeasts to develop, winemakers add little or no SO_2 to the grape must, which increases the risk of spoilage. Wild yeast species and higher bacterial development tend to produce greater variability and 'barnyardy' characteristics in the wine. The resulting wines may also contain higher levels of biogenic amines.

Alternatively, many winemakers can use dried yeast belonging to the *Saccharomyces cerevisiae* species. These are natural yeasts that have been multiplied in a controlled environment in a similar way to how plant nurseries propagate varieties of plants. *Saccharomyces cerevisiae* contains thousands of strains: hundreds have been isolated from natural ferments in wineries around the world (other strains are also used for brewing beer and breadmaking). Over the years, strains have been selected for specific traits such as tolerance to acidity, alcohol or for their ability to express particular varietal aromas: winemakers have quite a range to choose from. Ninety-five percent of the world's wine production is made with inoculated yeast.

A similar approach has made lactic acid bacteria strains available for winemakers to carry out the malolactic transformation in a timely manner. Specific strains can be co-inoculated at the same time as yeast, so that alcoholic and malolactic fermentations occur at the same time. This reduces energy use in the winery as there is then no requirement to heat the cellar to encourage LAB development, and wine stability is achieved earlier, limiting risk of spoilage as the wine matures. More recently, microbial research has focused on biocontrol products to replace the chemicals used to retain acidity, prevent oxidation and to reduce alcohol levels.

In addition to yeast and lactic acid bacteria, there are a variety of oenological products that may be used as yeast nutrients and processing aids to clarify and stabilize wine. Aside from SO_2 and allergens, these have not historically been listed on the label. But requirements are changing, and in the spirit of transparency this is a situation that should be embraced.

Sustainable wineries: energy

New Zealand's Felton Road, with a winery in Central Otago on South Island, is a model in its approach to energy. Since 2019 it has had a 32-kilowatt solar system that supplies all the winery's daily energy needs. But it has also greened its energy usage in other ways. The winery now uses only electric cars (plus one hybrid for heavy vineyard use) and has started using specialist electric vehicles in the vineyard. It runs a staff scheme to support the purchase of EVs and supply free recharging, which is available to customers and visitors too. Inside the winery, its efficient refrigeration system captures heat generated by the cooling system. The cellars are well insulated, reducing energy requirements, and the winery tries to use seasonal temperature changes as much as possible, such as opening the cellar doors in spring to warm the cellars to encourage malolactic fermentation.

It's not completely straightforward. 'Electricity is one of those nuanced issues. Is reducing electricity a good thing?' asks owner Nigel Greening. 'Our use has increased significantly because we now use 100 percent electric cars. Then moving to drones from tractors adds to the bill – but then again, with a reduction in diesel.'

Overall, winemaking accounts for 10–15 percent of a finished bottle of wine's carbon footprint – and 90 percent of winery emissions are due to the energy that wineries use. Electricity is often a winery's largest expense. The Australian Energy Market Operator (AEMO) estimates that around 40 percent of total winery expenditure in Australia is on electricity. Note that this is just the power that the winery uses to make wine: the equipment they use to do it, from tanks to barrels, also has a carbon footprint from its manufacture and transport.

The biggest consumer of electricity is refrigeration: according to the South Australian Wine Industry Association, this accounts for 50–70 percent of total power costs, although those are particularly high in Australia, which is both hot and where producers often keep storage tanks outside in the open air. But electricity is also used to run pumps, presses, filters, centrifuges, bottling lines and other winemaking equipment, as well as for air conditioning, heating and lighting. And as wine regions are generally located in warm or hot regions, wineries usually need to cool individual tanks, or the cellar with air conditioning, or both.

We discussed in Chapter 1 some of the options for wineries to reduce their greenhouse gas emissions by switching to renewable energy sources. But cutting electricity use is another important starting point for any producer trying to make wine more sustainably.

Monitoring electricity use – ideally in real time – and understanding energy use at each stage of the operation helps improve efficiency. Winemakers can reduce energy use by reviewing fermentation temperatures, must chilling and cold treatments. For example, harvesting at night, when ambient temperatures are at their lowest, reduces the need for cooling in the winery. Climate change is generally leading to grapes being picked earlier in the year, when external temperatures are higher, thus requiring more energy to reduce the temperature of incoming juice.

How much electricity a winemaker needs to cool their fermenting wine also depends on what it is stored in. Tank size, shape and construction material affect its thermal properties (as well as the amount of water required to clean it). Small tanks dissipate more heat than large tanks made from the same material due to the higher surface area to volume ratio. Stainless steel has a higher thermal conductivity, so it loses heat more quickly than oak or concrete, and is easier to clean.

The size, shape and construction material of fermentation and storage vessels have an impact on wine flavour and oxygen ingress, as well as determining the greenhouse gas emissions associated with manufacturing, shipping and installation, and the amount of water required for cleaning. Many quality-focused wineries aim to vinify grapes from individual plots separately in order to create wines that reflect the terroir of a specific site: to do this, they need to be able to harvest each plot at the optimum moment and ferment the grapes separately, often in smaller vats. Smaller vessels may require less energy to maintain a low temperature during fermentation, but generally require more water to clean. Stainless steel is energy intensive to produce but has a long life. Wooden tanks and smaller barrels of oak from managed forests are from a renewable source but are more difficult to clean, and barrels are often replaced after their third use. Recent trends have seen an increase of smaller concrete and clay amphorae, each of which have their own energy implications.

Cooling system equipment needs to be insulated and regularly maintained to ensure that it works at optimum efficiency. And when winemakers make decisions about buying and maintaining equipment, cutting energy use and increasing efficiency should be key factors. Meanwhile, using energy-efficient LEDs for lighting also reduces energy use.

The biggest capital investment most producers are likely to make is, of course, investing in a new winery – and at this key point, there are major opportunities to improve energy efficiency. Borja Gómez is head of architecture at IDOM Bilbao, a leading Spanish engineering consultancy. He led on the design of Beronia's new winery in Rioja, completed in 2020, which is Europe's first Leader in Energy Efficiency and Design (LEED)-accredited winery. It's a striking design, with 80 percent of the winery buried underground: it is almost invisible at a distance, from the main Logroño-Haro highway.

'We were pretty radical,' says Gómez. 'We realized that the temperature year round in barrel rooms was gradually rising because of climate change. The old winery design with the building above ground and barrels in the cellar underneath isn't working so well now.' IDOM's answer was a series of boreholes 100 metres deep, circulating cool water over the ceilings of the barrel rooms. This heat-exchange system also cools the main winery buildings in summer. 'At the same time', says Gómez, 'the whole geometry of the building is helping that – it's able to trap the heat of the sun when it needs it and avoid it when it doesn't.' The design has also brought huge savings in energy efficiency.

In addition, while most of the winery is below ground level, the 10-metre difference in height between the grape reception area and the winery floor allows some operations to be conducted by gravity, saving electricity that would have been used for pumps. Meanwhile, the winery is open to the light at the side, creating a more pleasant working environment. Indeed, IDOM regularly consulted the winery staff to get the design right. 'We worked closely with them – they were feeding into the process,' says Gómez. 'I had the feeling we were designing a kitchen.'

Southern Rhône producer Chêne Bleu has used gravity in this way by exploiting the steep, rocky topography of their winery's site. Winemaker Jean-Louis Gallucci explains: 'We were able to use the topography to create just what we wanted. We needed four levels, so we dug a hole 12 metres deep and nestled the top two storeys against a hill. It sounds easy but it was an enormous piece of work digging into solid rock. We used the stones they excavated as the base of the helicopter pad. We used special materials instead of the usual reinforced concrete.'

Sustainable wineries: saving water

Joe Uhr, director of winemaking at Gundlach Bundschu winery in Sonoma, California, looks out over an idyllic scene: a large lake, sparkling in the warm winter sunshine, bordered by Mediterranean trees and shrubs, with vineyards in the distance on the other side. Uhr points out a turtle sunning itself by the water below; he says there are crawfish, migratory birds and even otters sometimes. Yet this lake is, in fact, an artificial pond, part of Gundlach's sophisticated system for recycling wastewater from the winery.

Here, as in many regions, water is becoming a significant issue for the wine industry as the effects of climate change become starker. We discussed sustainable water use in the vineyard in Chapter 2. But even if they don't irrigate, wine producers use water at every stage of the winemaking process, above all (as we have seen) for cleaning everything, from tanks to pumps to floors, to prevent microbial contamination. All surfaces in direct contact with wine must be rinsed with potable water.

In South Africa, for example, Eben Olderwagene, environmental manager at Vergelegen, says they use around 85,000 litres a day in summer in the winery alone. Small Swartland producer Eben Sadie uses 5,000 litres of water a day; over in Stellenbosch, Meerlust, a medium-sized estate, uses more like 15,000 litres a day. And every litre of water used requires energy in its treatment and pumping. Using that much water creates a lot of waste: every litre of water used generates a litre of effluent. These factors can have significant environmental impact, and not just in terms of water supply: the greater the volume of dirty water a winery has to treat, the more energy it uses, which drives up its carbon footprint. Different cleaning procedures create cleaner or dirtier effluents, which can contain various organic and inorganic contaminants. These can have significant environmental impacts and need to be handled appropriately in order not to contaminate land and water flows.

California wineries are now legally required to clean wastewater before releasing it. Gundlach uses a dual pond system – one for wastewater piped down from the winery and a larger one for storing recycled water as well as collected rainwater. The winery adds bacteria that work to clean up the water, which are fed oxygen by aerators powered by floating solar panels. After up to a month of treatment, clean water is pumped up to the big pond, from where it's eventually used to irrigate vines in two of their sites.

A few hundred miles to the south, in Paso Robles, in California's Central Coast, Tablas Creek use an alternative system for cleaning wastewater. Tablas has created a wetland area close to the winery, with a settling zone to get rid of solids – such as grape waste washed off equipment – and then a gravel-bedded lake with water plants, which the water gradually feeds through. This not only provides a habitat for water birds and other wildlife, but is also a source of clean, recycled water to irrigate grass for their flock of sheep in the summer. Southern Rhône producer Chêne Bleu uses a similar approach: their wastewater goes to a half-hectare bamboo plantation behind the winery, where it passes through the plants' rhizomes. Naturally occurring microorganisms there degrade the organic matter, allowing pure water to be returned to nature. Warwick Estate, in South Africa, uses a technical version of the same process: a Membrane Bioreactor plant treats all the estate's effluent and sewage water, working biologically and through fine membrane filtration. The treated water then goes to a rainwater irrigation dam.

Capturing storm water is indeed one part of the water challenge, especially in places where climate change is making water conservation more urgent. In Catalonia, drought in recent years means that parts of the region now have a semi-arid climate, forcing some producers to bring in water for irrigation. 'It shows us what the future will be like,' says Sara Pérez, chief winemaker at Mas Martinet, in Priorat. 'Every time, the weather is more extreme.' Mas Martinet use pools to collect rainwater from roofs and others surfaces in autumn and winter: Pérez was very happy with the heavy rainfall they had in October and November.

Meanwhile, Warwick Estate has rerouted some roads and natural channels in their vineyards to redirect run-off into their irrigation dam. And in California, Jackson Family Wines collects around 1.9 million litres of rainwater each year, used mostly in refrigeration.

Wineries can also work to cut the amount of water they use in the first place. Mas Martinet, for example, now use steam cleaning at 150°C for barrels, which uses less water and does a better job than washing them down with power hoses. Likewise, Gundlach Bundschuh does the same for barrels and when switching over bottling lines.

Peak water consumption time is around harvest, when winery workers are constantly cleaning harvesting machines, hoppers, tanks, barrels, presses, pumps and hoses. Wineries can start by determining water-use hot spots to identify opportunities to reduce consumption. They can review their cleaning procedures for each task to make sure that they are doing an effective job with the least water possible. For instance, it helps to clean off all visible solid matter manually – grape skins, wine residues, tartrate deposits, yeast, pulp – before using water, as described by Rodrigo Serrano earlier in the chapter (*see* page 105).

Pre-rinsing or soaking can help with some equipment: this can be done with relatively clean recycled water. Then, depending on the cleaning operation, winery workers may use water with or without chemicals, filtered, at pressure, circulated, or sprayed onto equipment as a mousse. Pressurized hoses fitted with low-flow nozzles can work well and reduce the overall amount of water needed, even if they are spitting it out at 30–40 litres a minute. At Vergelegen, says Olderwagene, even their choice of paint for the winery floors makes a difference to how much water they need to clean off sticky deposits and the like.

Winery cleaning does involve the use of some chemicals and detergents to remove tartrate deposits and disinfect various surfaces, all residues of which need to be fully removed. Rinse tests can be used to confirm precisely how much water is needed to eliminate any chemical residues, without using too much. Relatively clean water can be captured and used for tasks such as cleaning floors. And if treated and filtered appropriately, wastewater can be reused for certain activities such as irrigation and flushing toilets.

Reducing carbon emissions: capturing carbon dioxide

In Familia Torres's Pacs del Penedès winery, large, silver, cylindrical balloons sit among the fermentation tanks, connected by metal pipes and slowly inflating. They look odd but serve a serious purpose: they are filling with the carbon dioxide produced during fermentation. Torres first trialled the system in 2015. When the balloons are filled, an air compressor is used to compress the gas to a pressure that allows it to be transferred to another tank for storage. The team then reuse it in the winery for inerting tanks to prevent oxidation. The winery now meets half of its CO_2 needs through this capture system.

Winemaking generates large amounts of carbon dioxide during fermentation – about 60 litres of CO_2 for every litre of still wine fermented, or around 125kg of CO_2 for every tonne of grapes. Making fortified wines produces even more. Most wineries release this CO_2 directly into the atmosphere, where it contributes to climate change: collectively, wineries produce in the region of three billion tons of CO_2

globally, although because it is biogenic in origin it is generally not taken into account in carbon footprint calculators.

Hence the Torres CO_2 capture project. Alcion in Bordeaux and Earthly Labs in the US have developed similar systems. Wineries use solid CO_2 – dry ice – and CO_2 gas on unfermented grape musts, as well as during storage, to protect grape juice and wine against oxidation. But CO_2 can also be sold: the largest users of CO_2 are carbonated drinks manufacturers, and it can also be used to produce sodium bicarbonate or potassium bicarbonate.

It should be noted that CO_2 is also a significant health and safety risk in wineries: winery workers die every year from CO_2 asphyxiation, such as the Ardèche vigneron found dead in one of his own tanks by his son in September 2024 in Saint-Montan, France. Carbon dioxide is heavier than air and tends to concentrate in pockets. At one million parts per million (ppm), the emissions from the top of fermenting wine tanks are the most concentrated CO_2 emissions of any industry in the world. (As a comparison, car exhaust fumes have a concentration of 200,000 ppm, compared to normal air at 420 ppm.) At 1,000–2,000 ppm CO_2 can cause drowsiness; above 5,000 ppm, it can cause oxygen deprivation; over 40,000 ppm is immediately harmful. Combined with warm alcohol fumes from fermenting tanks, it can offer up a deadly combination. Wineries need to be properly ventilated and test for CO_2 levels; some tanks also have guard rails at the top.

Cutting winery waste

At Domaine Bousquet, in Argentina's Uco Valley, the compost heaps are huge – long, low, dark mounds by the side of a series of vineyards, 823 tonnes in all in 2024. Bousquet farm organically, so they have significant compost needs for their vines. But their compost heaps are also a key method of cutting waste: much of the pomace from their winery – the solid waste of skins and pips – comes here to break down before being returned to the soil. Pomace used as compost on its own has higher potassium than is good for the vines. But when mixed, as here, with cow, sheep and goat manure, and with other organic waste such as tree leaves and straw, it is an ideal organic fertilizer. 'We follow the temperature and humidity evolution for months, to control and understand the phases of breakdown that we are in,' says Irma Remigio, environment and biodynamic management coordinator.

Grape pomace, in fact, is the largest contributor to waste generated by winemaking. Wineries generate up to 300kg of pomace for every tonne of grapes harvested, as much as half a kilogramme for every litre of wine produced. In total, the world's winemakers produce in the region of 13 million tonnes of pomace every year – a similar amount, for example, to the total global production of chickpeas. In Europe, disposal of pomace and yeast lees is regulated: wineries are legally required to dispose of these by-products appropriately. They are generally sent to a distillery, where tartaric acid and ethanol are recovered to be used in various industrial processes. France recycles 97 percent of organic winemaking waste in this way, with distilleries processing 850,000 tonnes of pomace and 1.4 million hectolitres of wine lees each year. Yet in California, grape pomace is often left to decompose in the sun; similarly, many other New World wineries fail to consider ways in which these by-products can be appropriately disposed of.

But pomace is useful: as well as in composting, it can be made into biochar or turned into energy feedstock. And it has more complex industrial uses thanks to its wealth of component chemical compounds. In Australia's Barossa Valley, for example, Yalumba, part of Hill-Smith Family Estates, sends its pomace to be processed by Tarac Technologies, which extracts any remaining alcohol and produces other grape-derived products such as natural colourings. Pomace is also increasingly being seen as a valuable source of phenolic compounds and polysaccharides for the cosmetics and food industries. Added to which, phenolic compounds are being tested for their use in drug treatments to combat cancer, diabetes and other diseases. Pomace is a source of tartaric acid, which is widely used in the food and pharmaceutical industries. Domaine Bousquet sends some of its pomace to the tartaric acid industry in this way: 'They extract the tartaric that we need to use in the next season,' says Remigio. Studies have also suggested a role for grape pomace as a biofuel, either for anaerobic digestion to produce biogas or for bioethanol.

Wineries can approach their wider problem by starting with a waste audit, to understand the kind of waste they produce and how much there is of it. Sustainability experts talk about the 'waste hierarchy' – ranking methods for disposing of waste in order of desirability for the environment. Preventing waste is the best option, followed by reuse; landfill is the least desirable. Wineries can then develop systems to sort waste streams, to separate general waste from dry recyclable materials in appropriate bins. And they can help reduce waste further up supply chains by, for instance, choosing supplies packaged with recycled materials.

After pomace, wine filtration processes also generate waste in the form of filtered wine lees, used filters and filtration aids such as perlite, an inert powder made from volcanic rock. Yalumba sends used perlite and lees from filters for processing, where potassium bitartrate (Cream of Tartar – used in baking and some household goods) can be extracted from them.

Wineries, of course, also generate waste in much the same way as other businesses, principally in packaging. To be more sustainable, they need to separate waste streams and recycle all they can. For example, at Tetramythos winery in the Peloponnese, there are separate trash and recycling bins for different materials, sited just outside the winery, so that local villagers can recycle waste there too. At Spier, in Stellenbosch, South Africa, international sales manager, Pierre Nortje, says that all new staff members go through a two-day induction course that looks at how the company does business, including a full briefing on its sustainability programme. One of the topics is making staff aware of packaging choices: 'We are encouraged not to bring our lunch meals in single-use plastic,' says Nortje. Staff are also taken to Spier's waste-sorting facility to create awareness of what happens to the things that they throw away. This has helped Spier achieve 98 percent recycling of its waste.

Some wine producers generate significant numbers of used oak barrels, often discarded after they have been used three times, having thereby lost much of the flavour they impart to wine. Using oak barrels itself adds to wine's carbon footprint because of the energy used in their manufacture and transport. Long-running debates rage in the wine world about the desirability of flavours imparted by oak barrels, but their use does have an added environmental cost: one study in Rioja found that using 225-litre barrels has an impact of more than 3kg of CO_2 equivalent per litre of wine every six months. Meanwhile, at Yalumba, old barrels are repurposed for storage or made into furniture.

Key lessons for sustainable winemaking

› **Adopt a circular approach**: reuse waste products where possible, such as using grape pomace for composting, retrieval of tartaric acid and recycling packaging.

› **Minimize energy consumption** and make sure that what you do use is generated from renewable sources.

› **Minimize water usage** by cutting your consumption where possible; review cleaning procedures and recycle winery water. Capture rainwater for use in the winery and elsewhere.

› **Treat waste water** to minimize environmental impact.

› **Minimize winemaking chemical inputs** such as SO_2, and label what you use transparently.

› **Practise zero waste to landfill** by cutting organic and inorganic waste and by recycling.

› **Capture CO_2 produced from alcoholic fermentation**.

Chapter **FOUR**

BOTTLING AND PACKAGING

At Ardagh Glass Packaging's sprawling factory in Barnsley, South Yorkshire, UK, the noise when you stand next to the bottle-moulding line is deafening. On screens, in a room somewhere above the conveyors, a video flickers of the interior of the furnace, a chamber filled over a metre deep in molten glass. Bright orange 'gobs' of glass are cut from a stream emerging from the fiery depths, which then shoot down to the moulding machines below. A row of 10 robotic arms lift the red-hot wine bottles – they are still at around 550°C – from the moulds two at a time, every eight seconds, and put them on a conveyor belt that travels down the production line. This furnace can turn out 345,000 bottles a day.

People who care about wine, or write about it, tend to concentrate on what's in the bottle and in their glass. Yet packaging is a vital aspect of wine, especially when it comes to sustainability. Packaging has obvious functional roles in protecting a wine's quality and ensuring that it reaches consumers' glasses in tip-top condition. It must provide mandatory information, too – for instance, the alcohol content on the label. But packaging also has an aesthetic role and can add value to the overall product, over and above the value of the wine it contains. It is an important part of attracting wine customers through design, creating desirability, and differentiating between wines. And of course, wine packaging has a cost. For lower-priced wines, packaging is a significant proportion of the total production cost – indeed, in some cases, the cost of packaging outweighs the cost of the wine therein. And for all wines, damage and poor wine quality creates waste and can lead to dissatisfied and ultimately lost customers.

As we have seen, making wine involves myriad environmental impacts. None, however, are as great as those that surround the vessel in which most wine is carried from its source to your table: the bottle. Globally, around 85 percent of wine is sold in glass bottles. Wine packaging and transport contributes up to 50 percent of wine's carbon footprint. But by any measure, packaging involves more emissions than all the electricity used in the winery and all the fuel used in transporting the wine put together.

Yet wine's packaging is evolving in terms of the materials it is made from, its shape and its image. And wine consumers are diverse too, consuming it at a variety of social drinking occasions: indoor and outdoor, private and public events, concerts, festivals, sporting events, at home, in the garden or at more formal meals. The traditional glass bottle may not be the most appropriate packaging container for all occasions. Indeed, the use of glass is prohibited in places for health and safety reasons. Meanwhile, a significant amount of wine is served in restaurants and bars: the packaging needs of trade customers are different. And new forms of packaging may attract new, younger wine consumers at a time when global wine consumption is in decline. In the US, for example, sales of wine in aluminium cans increased 30 percent between 2012 and 2018, with a higher proportion of cans sold to 21–34 year-olds. One 2024 study predicted that canned wine sales will grow globally by 17 percent a year over the next decade.

As with glass bottles, other packaging can have a big impact on a wine's carbon footprint. But it also has an impact on the environment in other ways. This is assessed with Life Cycle Analysis (LCA), a standard method used to evaluate the total environmental impact of a product during all stages of its life, from the extraction of natural resources to final waste processing. Thus, some packaging materials require large amounts of energy to extract and process, leading to high carbon footprints. And the weight of materials also affects greenhouse gas emissions generated by transportation.

But after a product has been consumed, what happens to its packaging also contributes to its environmental impact. Packaging makes up a significant proportion of collected household waste. Recycling rates for consumer packaging globally are less than 15 percent overall: they vary dramatically according to the packaging materials and where they're being thrown away, but the majority goes to landfill or is incinerated. Some packaging waste is dumped and ends up as litter on land or in the sea: it is estimated that, overall, 41 percent of packaging is mismanaged, and of that percentage almost half ends up in the ocean. And depending on what it is made from, as packaging degrades, it can release methane, toxins or other pollutants.

Governments internationally are increasingly shifting the burden of the cost of managing packaging waste to producers and retailers through the introduction of taxes such as the UK's Extended Producer Responsibility (EPR), which partly or fully funds the cost of collecting, sorting and recycling. EPR schemes exist in many places and are growing in ambition. The UK's Wine Society estimates that the British EPR starting from October 2025 will cost it an additional £1.3 million in 2025 – and this is a business that predicted a profit of just £2 million in 2024.

There has never been such a wide variety of wine packaging formats, with different materials, shapes, weights and colours available, including bottles made of glass, round and flat PET (polyethylene terephthalate) bottles, so called 'paper' bottles sealed with a variety of stoppers, or screw cap, aluminium cans, bag-in-box, pouches and cartons. Each format offers its own set of consumer benefits and technical challenges: shelf life, cost, recyclability and carbon footprint. Some require investment in specialist filling equipment, while packing lines have high minimum runs and costs to fill. Meanwhile, composite materials – for example, where a container has a metallized inner – are usually more difficult to separate after use and correspondingly incur low rates of recycling. This chapter explores the current options available for wine and the pros and cons of those options.

Packaging principles

The broader principle here is the goal of a circular economy. Such an economy minimizes the amount of packaging used, reuses it or makes it from recycled materials, and recycles it after use, sending zero waste to landfill or for incineration. Minimizing packaging can reduce costs and drive logistical efficiencies, at the same time as reducing greenhouse gas emissions from both the packaging's manufacture and transport. The ultimate aim of a circular economy is a closed-loop system with no use of virgin materials. And the more times packaging is used, the more efficiently the resources and energy that went into making it are utilized.

Some materials, such as glass and aluminium, can be recycled an infinite number of times without losing their properties. Others can be recycled a certain number of times before impurities affect their performance or require a proportion of virgin material for strength. Thus, in reality, the loop will never be completely closed: there is always some loss of material in the collection and recycling processes. Meanwhile, demand is not static – so where demand for a certain product is rising, that will require more virgin materials.

Packaging materials fall into two broad categories: 'technical' and 'biodegradable'. The technical category, made up of packaging materials that don't biodegrade, includes metals, plastics and glass. These are manufactured from non-renewable resources. Extracting raw materials, such as the mining of bauxite, causes environmental damage, while smelting aluminium from bauxite requires high amounts of energy. The most efficient use of technical materials is to recover and reuse them, as less energy is generally required to recycle them than to produce packaging from virgin materials. This also keeps finite resources in circulation and limits the amount of waste to landfill. But not all are recyclable. Most plastics are produced from fossil fuels; they are not recyclable, do not biodegrade and do not break down for hundreds of years.

Biodegradable materials such as paper, cardboard and wood can be broken down by microorganisms as part of a biological process. Cardboard is widely used to protect goods during transport and distribution: it makes up more than two-thirds of all packaging. However, as we saw in Chapter 2's section on the carbon cycle, there are quite large differences in the time it takes different materials to biodegrade. Compostability implies breakdown within a relatively short time scale – but some materials identified as compostable are only broken down in industrial facilities which operate at higher controlled temperatures (50–60°C) to speed up the process, and cannot be composted at home.

Recycling

Inside a vast hangar in Bermondsey, south London, a bewildering series of conveyor belts are running up and down in different directions, like something out of an MC Escher drawing. This is the Materials Recovery Facility at the London Borough of Southwark's huge Household Refuse and Recycling Centre. The din of the machinery is deafening – visitors must wear industrial ear defenders – and the smell is insistent. Trucks filled with the contents of local households' recycling bins roll into a massive reception area: this plant, run by giant French multinational Veolia, handles waste not just from Southwark but also from the boroughs of Westminster and Greenwich, and the City of London area in the financial district, as well as from the nearby municipalities to the east of Medway and Southend. Of the waste that goes in, just under a third is recycled and just over half goes for incineration at a nearby plant that will extract heat and power from the process. Less than one percent is goes into landfill.

Inside, the Southwark plant runs 24 hours a day, seven days a week. First, the recycling goes through a pre-sort cabin where workers attempt to remove larger or potentially dangerous items by hand. Then it is blown over rotating discs to separate smaller and larger materials: glass and smaller bits of paper fall through to be carried away on other belts. There is some infrared sorting of the mixture of paper, cardboard and plastic that comes out, and then it is hand-sorted by rows of workers in face masks and ear defenders, getting rid of plastic bags, plastic film and other non-recyclable bits. Further on, steel items are removed by magnets and aluminium cans by an eddy current separator. A robot picks out remaining non-recyclable rubbish from this belt, before the cans go to be packed in bales for onward transport to processors. The same happens to paper and card, and to PET plastic bottles.

To watch the process in action is to understand the enormous complexity and challenges of recycling. Moreover, this is just part of the process: much depends on how the materials are then processed. Southwark does not sort bottles by colour and sends off a mixture to processors: it is all 'downcycled' for use in construction. Some glass processors use optical sorters to separate the glass into different colours, so that the crushed cullet (glass in its broken-down state) can then be recycled as bottles. But, overall in the UK, just over a quarter of recycled bottles go to construction aggregate.

The quality of recycling materials and how they can be used thus depends a lot on whether materials are segregated or mixed, collected kerbside or from central points. These factors have a big impact on actual recycling rates – and vary a lot by country and region. It is no coincidence that the UK's highest recycling rate is in Wales, which, in fact, boasts one of the highest rates in the world at 66 percent (2024), because householders have to segregate their recycling by type of waste. Even here, there are variations between areas – though the overall figure is a dramatic improvement on the Welsh rate back in 1997 of just nine percent. In addition, to the annoyance of the British glass industry, around a quarter of the UK-processed cullet is exported internationally: in total, just 36 percent of bottles recycled in the UK go to make new bottles there.

'We are campaigning very hard to get this figure increased and have launched a campaign called Close the Glass Loop to work mainly with local authorities to get much more glass back into our furnaces,' says Matthew Kay, glass packaging policy lead at industry body British Glass. 'Firstly, we need the quantity of it to increase and then we need the quality, so ideally glass should be collected separately like it is in large parts of Wales. We need investments in technology and then we need to stop good-quality stuff being exported.'

Collection schemes for glass and aluminium exist in most places, with varying degrees of uptake. Plastic packaging accounts for over half of collected plastic waste in Europe, and while plastics recycling is growing because of environmental concerns around plastic pollution, it represented just six percent of total plastics production in 2019.

Plastics are widely used for packaging because they are lightweight and durable. But while some plastics such as PET can be recycled, a large proportion cannot.

Soft drinks companies are already switching from PET bottles made from virgin material to alternatives made from 100-percent recycled PET (rPET). This switch creates a market for recycled material, encouraging more to be collected and processed. PET is the most widely recycled plastic, though recycling rates still remain low. The sale of goods in single-use plastic, including drinks containers, is being banned in many regions by regulations such as the EU Single-Use Plastics Directive of 2019. This sets a target of 77 percent collection of plastic bottles with caps and lids by 2025, and 90 percent by 2029. The EU also has a target of 25 percent recycled content for PET bottles by 2025, though at present demand for food-grade rPET for bottle production exceeds supply. An increasing number of EU countries are also adopting Deposit Return Schemes for PET bottles.

The detail of how such recycling schemes work might seem mundane, but it is important. True recycling, creating a closed-loop system for the materials involved, requires effective collection schemes, where packaging materials are easily separable and retrievable, and where there is a demand for the material in the same region in which it is disposed. It also needs to make sense economically: the value of the recycled material obviously needs to be higher than the costs of collecting and processing it. So, low-value materials and the cost of transport or lack of recycling infrastructure mean that a significant proportion of recyclable collected materials worldwide still end up in landfill. The low cost of plastic, for example, is one of the reasons there has been little incentive to recycle: it has been cheaper to produce more using virgin material. Processes to chemically recycle plastics into oils and liquids equivalent to virgin material, to create an infinite loop, are in development but some way off being commercially viable. But complex multilayered technical materials, such as the mixed plastics and metals used in bag-in-box, pouches and Tetra Pak, are difficult to separate and generally don't get recycled.

The discrepancy between where products are made and where they are consumed is also problematic. In the case of the global wine trade, the main wine-producing countries are net exporters of packaged wine, while consumer markets, which import coloured glass wine bottles (such as the UK) have low domestic demand for green glass cullet, although the increase in UK-bottled wine is going some way to increasing that demand. There is also some loss between what is collected and what is fit to go for recycling, due to contamination and other issues.

So the choice faced by wine producers and brand managers as to what packaging to use is an important one: technical requirements, consumer needs, cost and sustainability factors all need to be taken into account. All this means that wine is trickier to package than you might imagine. Its flavour changes over time. All wines have an optimum drinking life, which can vary from around 12 months to decades after filling, and depends on the wine's quality, style and packaging. While in the UK, 92 percent of wine is consumed within 48 hours of purchase, lead times for global distribution and retail rates of sale often require a keeping life of 12+ months, to avoid waste.

Packaging in contact with wine – its 'primary packaging' – must be made from materials suitable for food. Wine is sensitive to oxygen, ultraviolet light and temperature. Wine packaging helps protect against these and prevent any ingress or contamination from gases, odours and microorganisms. And because wine is high in acidity and contains alcohol and sulphur dioxide, wine packaging must be chemically resistant to those things. Packaging materials vary in terms of their oxygen barrier properties and reactivity with wine, with an impact on shelf life. But glass is the only material suitable for sparkling wines and wines drunk 18 months after filling. For sparkling wine, the packaging must also be able to withstand five to six atmospheres of pressure, or 70–90 pounds per square inch, two to three times the pressure of a car tyre.

'Secondary packaging' is the cardboard cartons or wooden boxes that protect bottles (or other packaging formats) while in transit. Cardboard is widely used to protect goods during transport and distribution: it makes up more than two-thirds of packaging. The strength and design of secondary packaging depends on the primary packaging it contains. It must be appropriate to the supply chain it will go through: a carton to protect glass bottles stacked on a pallet requires a different strength than one for aluminium cans or bag-in-box wine. Flexible primary packaging such as pouches need more support from an outer case; and an outer case delivered by a third-party logistics provider to the end consumer needs to be stronger than for retail distribution. Secondary packaging includes tape, adhesive to seal cartons, packing materials, case dividers and outer carton labels.

'Tertiary packaging' is used to protect products when they are shipped in quantity, holding cases together in a stable configuration to prevent damage and increase efficient use of space. Wine cases are often stacked on a wooden pallet or 'slip sheet' – a thin, pallet-sized sheet of plastic, paperboard or fibreboard – with strapping and shrink wrap used to hold them in place. Tertiary packaging can include layers made from paper or cardboard, which prevent slippage and help stabilize pallets, plastic film or plate, or wood, and plastic wrap. Tertiary packaging is functional, does not usually reach the end consumer, and is disposed of by the warehouse or retailer.

Packaging's shape determines the stability of a stacked pallet and how efficiently space is used. For instance, a lot of space is lost when round bottle shapes are packed together, whereas flat and cube shapes efficiently fill 100 percent of the space, driving logistics efficiencies. Sustainable packaging requires a holistic approach to the entire package.

Plastics are used in primary, secondary and tertiary wine packaging in pouches and bags, synthetic stoppers, capsules, plastic screw caps, aluminium can coatings, screw cap liners, shrink wrap, adhesives, tape, cork treatments and bottle coatings.

Many large spirits and soft drinks companies are now reducing packaging. Treasury Wine Estates (TWE), the Australian international drinks company, aims for 100 percent of its product packaging to be recyclable, reusable or compostable, originally by the end of 2022.

TWE's Sustainability report in 2024 shows some of the complexity in addressing sustainable packaging with 13 ongoing projects. Meanwhile, Diageo, one of the biggest drinks multinationals, is phasing out 183 million cardboard gift boxes from its premium Scotch whisky portfolio.

Glass bottles

Outside, it is a raw, foggy January day at Encirc's vast plant in Elton, Cheshire. On the factory floor, however, it's warm, the air heated by the mammoth machines and the molten glass coming out of one the two glass furnaces, the largest in the world. To the side of the furnace, an automatic system feeds cullet – recycled glass crushed into granules of about a millimetre, with an admixture of sand – into the towering machinery. Between them, these furnaces make 2.5 million tonnes of glass a year, much of it from recycled bottles. And far down the production line, beyond the areas where the finished bottles are treated and stacked, there is an enormous bottling area where they are filled with bulk wine shipped in from South Africa, Australia and Chile (we will return to this in the next chapter). The bottles coming full circle at Encirc illustrate many of the challenges of this most traditional and iconic form of wine packaging.

Every year, the global wine industry uses the equivalent of around 27 billion 75cl glass bottles. Glass bottles have been used for wine for centuries. A bottle will always carry more prestige than a can or box. Glass is inert, impermeable and a barrier to oxygen. Coloured glass provides some protection against UV light. Amber glass is the most efficient at this, although clear glass bottles, known as 'flint', widely used for rosé wines, provide no UV protection, which can lead to rosé having off characters as a result of 'light strike', gained after sitting overly long on retail shelves.

Glass can be washed, sterilized and reused – although most wine bottles today are single use. Glass is also highly recyclable, retaining its intrinsic properties infinitely no matter how many times it is recycled, though it does need a small proportion of virgin materials added to the recycled cullet to maintain quality. Green bottles can be made from up to 96 percent recycled glass, with a little sand added for the balance; for the clear bottles used for rosé and some whites, the recycled content is only 30–40 percent, as more than that introduces visual impurities into the finished product. As mentioned, glass can also be crushed and 'downcycled' as road-building fill.

But glass bottles are the single biggest source of carbon emissions in the wine industry. Bottles' manufacture alone accounts for at least 20 percent of the average bottle of wine's carbon footprint – as against about 15 percent for growing the grapes and making the wine. Taking into account the weight of glass to transport, bottles are responsible for at least 30 percent and up to 50 percent, depending exactly where and how they were made and how far they were transported to be filled.

Whether the glass is being made from virgin materials – sand, limestone and soda ash – or recycled cullet, its components need to be heated in a furnace to between 1,500 and 2,000°C to melt into liquid (cullet melts at a couple of hundred degrees lower than sand). And since the late 19th century, the normal furnace fuel has been gas. The International Energy Agency estimates that container and flat glass industries emit over 60 million tonnes of CO_2 per year. An efficient glass furnace requires 1,100 kilowatt-hours (kWh) of energy to melt a tonne of virgin glass – roughly the amount an average home uses in a whole month. In addition, soda ash and lime carbonates give off CO_2 when melted with silica sand. Altogether, a furnace producing 300 tonnes a day consumes around 32,000 cubic metres of natural gas and releases 62 tonnes of CO_2 each day. It adds up to perhaps as much as 10.4 million tonnes of CO_2 a year.

Not only that, but a glass furnace can't be switched off. It takes too long to get up to heat, and molton glass in the system can't be allowed to cool and set. So once on, the furnace runs continuously – albeit not always at full power – over its 15-plus year lifetime. For these reasons, 80 percent of the carbon emissions in glass manufacture are associated with the operation of the furnaces.

One of the wine industry's easy wins in reducing emissions should be to cut the weight of wine bottles. The heavier the bottle, the more raw materials and energy are needed to produce it. And the more fuel that is used to ship heavier bottles, the higher their carbon footprint. Using lighter-weight bottles immediately reduces a wine's carbon footprint: replacing a one-kilogram bottle with another weighing 400g reduces CO_2 emissions by 231g per bottle, or 60 percent. And there should be significant cost savings considering the huge increase in glass costs in recent years, with a jump of roughly 40 percent during the energy price inflation crisis of 2022–23 caused by Russia's invasion of Ukraine. Using lighter-weight glass bottles also reduces potential Extended Producer Responsibility costs, where such controls are in place. In the UK, these are based on tonnage, with a projected cost when the scheme starts in autumn 2025 of £240/tonne of glass.

Yet some bottles can still weigh up to 1.2kg – not including the wine they contain. Heavy bottles are an entirely unnecessary and wasteful marketing choice, unfortunately associated by importers and consumers in some markets with higher-quality wine. 'Reducing bottle weights increases your chances of commercial success in mature markets like the UK and northern Europe,' says Rafael de Haan, co-owner of Herència Altés, in Terra Alta, Spain. 'The US is more open, too, which wasn't the case five years ago. But there are still hurdles to overcome selling prestige wine in lighter bottles in less mature markets like China and Latin America.' Nigel Greening, owner of New Zealand's Felton Road, disputes this. Despite his wines' 390g bottles, he says: 'There's never been an issue, including in China. It's a myth as far as I can see.' He points out that one of California's most sought-after wines, Screaming Eagle, comes in a standard 550g Bordeaux bottle.

Valentina Lira, sustainability director at giant Chilean producer Concha y Toro, says that their average still-wine bottle weighs 450g, and they have committed to lowering that to 420g by the end of 2026. But she insists: 'In the Americas, people are very attached to the idea that a heavy bottle means a good wine.' They shouldn't be. 'One-kilogram bottles? We could get three out of one of those!' jokes a manager at Ardagh. It's true: the company makes 400g and 300g bottles. In fact, Ardagh has a more radical research and development project that is trying to develop coatings that would allow the glass to be even thinner, yet keep the same strength. But those seem to be some way off commercial production.

And finally, where bottles are made, where wines are bottled, and the destination market the wines are sold in have an impact on packaging and distribution emissions. In Europe, over 40 percent of glass bottles are transported more than 300km just to reach their filling destination. And there are many wineries in the US and South Africa sourcing glass bottles produced in China. After filling, bottles are often transported long distances within domestic markets, while nearly 15 billion glass bottles are shipped across international borders every year. We examine this in Chapter 5.

There aren't yet many wine bottles on supermarket shelves weighing in at under 400g. The Sustainable Wine Roundtable's Bottle Weight Accord has set the goal of reducing the average weight of 75cl still wine bottles to below 420g by the end of 2026. The shape and design of a bottle influences how light it can be made. At one stage, it was generally easier to make lightweight, Burgundy-shape bottles than straight-sided ones like those used in Bordeaux, but this is no longer the case.

The more knocks a glass bottle receives, the more fragile it becomes. But the biggest issue in getting producer acceptance of 300g bottles, say Ardagh, is shock resistance on bottling lines, where bottles bump into each other and there is some potential for metal-on-glass contact with machinery. Ardagh are working with manufacturers of lines such as Germany's Krones and big bottlers such as Greencroft, near Durham, UK, using equipment such as shock loggers and high-speed cameras to understand exactly the pressures bottles are subjected to.

It's trickier to reduce bottle weight with bottle-fermented sparkling wines such as champagne, which need more robust bottles to withstand internal pressure of up to 5–6 bars; the actual pressure capacity of champagne bottles is considerably higher than that. There has been some reduction in the weight of champagne bottles, and the authorities have worked with wine producers and glass manufacturers to reduce it further, though, for the moment, those efforts appear to have ground to a halt around the 835g mark. Champagne Telmont uses an 800g bottle. Spanish Cava and French sparkling *crémant* wines, where bottles have an internal pressure of 3–4 bars, can be 775g. Tank-fermented sparkling wines such as Prosecco can use lighter 650g bottles. But part of the concern in reducing sparkling wine bottles' weight is over safety.

Bottle lightweighting is now part of the strategy at all big producers with sustainability commitments. For example, in South Africa, both Boschendal and Spier, in Stellenbosch, have bottle lightweighting targets built into their sustainability strategies. But Iona and Radford Dale, in nearby Elgin, small, but fleet-footed organic or biodynamic producers without sustainability directors, are both already using lighter bottles than those big producers.

Another option is to reuse bottles. On a nondescript industrial estate in northeast London, Sustainable Wine Solutions (SWS) is a hive of activity as workers fill 20.5-litre, stainless-steel kegs from 1,000-litre international bulk containers and move them around the warehouse. Up to 65 percent of the 110,000 litres-plus that SWS import annually is sold in this way, through bars and restaurants. The kegs use the same gas technology as beer kegs, and last around 30 years, says Calvin Pearson, SWS's head of wine: 'You're more likely to lose them than have them suffer damage.' SWS also deliver other types of kegs and bottled wines and collect empties, reusing bottles up to 30 times.

Such refill schemes do, however, depend on a closed-loop system where a single operator collects and refills the bottles. In 2022, UK grocer Tesco did a trial with Loop and Terracycle, which included wines sold in reusable bottles with a refundable 50p deposit. Such schemes can be operated in part alongside deposit return schemes designed to encourage consumers to recycle packaging: in Germany, for example, consumers pay a deposit on any bottles or cans, which they are refunded when they return the empties to an in-store return machine. However, greenhouse gases are generated in the collection, washing and refilling of bottles, which needs to be factored in.

'Reuse is better, if possible,' says Josep Maria Ribas, climate change manager at Torres, in Catalonia. Torres trialled a reuse scheme in Scandinavia, where in-market bottling and state-monopoly stores made it easier. Torres also conducted a scheme where they shared a 430g bottle design with fellow Spanish producer González Byass, taking advantage of the fact that one of the few washing facilities in Spain is close to their winery in Penedès. This gave an indication of how difficult it is to use refill schemes at scale: it requires substantial infrastructure with customers returning undamaged bottles, space for storage, collection and transport to a central site for cleaning and refilling. And the bottles can only be reused a certain number of times before becoming scratched, if not chipped or cracked. Ribas says they found that in terms of what was acceptable to customers regarding appearance, up to 10 return/wash/refill cycles was the limit. Also that for a bottle reuse scheme to be environmentally and economically viable, the reverse logistics distance must be less than 220km, and there must be a minimum of seven to eight reuses.

That leaves the recycling option. In the UK, glass cullet is produced by a small group of large, specialist processors – notably URM, Recresco and Veolia – who take recycling waste collected by local councils. Remaining bits of capsules on bottles are blown off, while labels float off when the bottles are soaked. Metal caps, or their remains, are separated out too. Robin Thompson of Encirc says that it's better for consumers to leave screw caps on bottles, from where they can be recovered and recycled, than try to recycle them separately: caps can get stuck in machinery. Processors must also separate out non-glass materials left in recycling: ceramics are a particular problem if they haven't been taken out, sometimes ending up in lumps in bottles, rendering them unusable.

The end product of bottle recycling is furnace-ready glass cullet, a product which has soared in price in recent years: Thompson says that it would now be cheaper for Encirc to work using virgin materials. Critics says that in the UK, not enough of the funds generated by recycling go back to local government, which does most of the collection. In theory, this situation should improve with the advent of Extended Producer Responsibility for recycled materials, where, from October 2025, manufacturers will pay levies on consumer waste they generate, money from which will end up with councils to help cover collection costs.

It takes 30 percent less energy to melt a tonne of recycled glass cullet and there are no CO_2 emissions from soda ash and lime carbonates either. Going from a 1kg virgin glass bottle to a 400g bottle produced from 80 percent recycled glass reduces emissions by about 70 percent. But such figures depend in part on the effectiveness of recycling collection schemes. In Europe, container glass has been collected and recycled for over 40 years: the current recycling rate across the EU is 76 percent. In the US, it is just 33 percent (2019) and in Australia 46 percent. But the actual amount of glass that goes back into producing new bottles varies from 77 percent in Germany to 36 percent in the UK and only 21 percent in the US. Close the Glass Loop aims to boost recycling collection rates to 90 percent by 2030. Sorting and processing techniques are improving.

The long-term solution to making glass more sustainable is to cut its manufacturing emissions. This is technically challenging. Ardagh run more efficient furnaces in various places such as Doncaster, South Yorkshire, that can save up to 14 percent in emissions. Major international glass group Verallia, the world's third-largest producer of glass packaging, introduced the first all-electric glass furnace in 2023 at its plant near Cognac, France. It says that this reduces carbon emissions in glass manufacture by 60 percent.

In the same year, Ardagh unveiled its NextGen furnace at Obernkirchen, Germany, which will ultimately operate on a mixture of 80 percent electricity and 20 percent natural gas. Ardagh says this will lead to a 69 percent drop in emissions. The next step will be to replace the remaining gas with hydrogen. Ardagh have one Hydrogen Electrolyzer plant in Sweden which replaces up to 20 percent of the natural gas used with 'green' hydrogen made on site from renewable energy. They intend to combine this technology with an electric furnace in the future. Encirc in the UK, part of the Verallia group, is exploring hydrogen-powered furnaces. It is also working in collaboration with Diageo on a waste-based, biofuel-powered furnace.

But, realistically, the timeline for widespread implementation of these projects is some years in the future.

This is partly because it's all very expensive: investment in retooling the Verallia Cognac plant cost €57 million (US $66 million), while Obernkirchen took Ardagh six months to build: it costs more both to build and run than a conventional gas furnace. Electric furnaces are expected to have shorter lifespans than gas installations (around 10 years rather than 15 – though they're too new for any to have been decommissioned yet). The plants also demand a consistency of renewable grid power supply that is not yet available in many countries.

Bottle closures

Whatever the type of bottle, it needs a closure: for centuries, this has traditionally been a cork. Bottle closures are a relatively small component of a wine's primary packaging – just one to two percent of the total – but the seal they give is vital for wine quality. The quality of the closure and the amount of oxygen that can pass through it have an impact on maturation and drinking quality. Tiny quantities of oxygen enter the bottle through a cork or screw cap, measured as the oxygen transmission rate. Closures can vary from offering a completely hermetic seal, to allowing rates of up to 20mg of oxygen ingress a year. Closures protect wine from contamination, should be tamper-proof and easy to use, and must be acceptable to consumers.

There are two main types of bottle closures – stoppers and screw caps. Stoppers – usually cork – are compressed and inserted into the neck of a bottle. The elasticity of the stopper causes it to spring back to shape, exerting pressure against the neck of the bottle and creating a tight seal that extends the length of the inserted stopper. Screw caps consist of several component parts with an aluminium or plastic body and plastic wadding inside. They sit over the top of the bottle neck and create a seal between a 'liner' cushioning the inside of the cap and the lip of the bottle. The seal is created by applying a downward pressure to the top of the screw cap and moulding the body of the cap to the thread on the bottle neck using rollers. Screw caps have grown in popularity thanks to their ease of opening and resealing, and perceptions associating them with cheap wine are gradually disappearing. The small size of closures poses challenges when it comes to recycling, as they often go undetected and fall through sorting conveyor belts. Actual recycle rates for bottle closures are very low and the majority go to landfill.

Cork

Carlos Abreu is standing next to a young cork oak tree (*Quercus suber*), explaining its life cycle. Around its thin trunk the virgin bark is knobbly and never harvested. That first harvest will come when the trunk reaches a circumference of 70cm at chest height, at an age of 20 to 25 years. After a further nine years, its regrown bark will be harvested again. And after another nine years (around 40 years since the tree was planted), when the bark is growing back more evenly and smoothly, it will give up its first useable cork. Over its 200-year life, the cork tree will be harvested around 17 times. It takes patience. As Abreu says: 'Imagine going to an investment company and saying, "Guys, I've got a plan to make money in 40 years!"'

The cork oak grows throughout the Mediterranean basin – though Portugal produces approximately half of all cork harvested each year and around 70 percent of the world's wine corks. However, natural regeneration of the cork forest is limited – many landowners graze animals that eat the acorns – and nobody has been planting much since the British port trade planted new forests in the early 20th century. But this is the cork business and it follows the logic of Amorim, the owner of this plantation in Rio Frio and the world's largest cork producer.

Cork is by far the most sustainable way of sealing wine bottles. It is lightweight, compressible, elastic, impermeable, buoyant, fire-retardant and durable. Analysis by Price Waterhouse Coopers (PwC) shows that the environmental impact of producing aluminium and plastic closures is significantly higher than that of cork. Cork is a natural product that also locks up carbon (until it is burned or decomposes): the Mediterranean cork forest from which it originates is a natural carbon sink, sequestering 14 million tonnes of CO_2. In fact, analysis by KPMG and Ernst & Young has shown that cork has a *negative* net carbon footprint: the average of 2.1g of greenhouse gas emissions per cork from its manufacture is offset by its 7.05g value of stored carbon. And if you take into account the carbon permanently stored in the forest, then each cork stopper has a net carbon value of -276g.

Cork oak forests in Portugal have been protected by law since the 13th century, and now many of them are Forest Stewardship Council (FSC) accredited. The cork oak forests support biodiversity and provide a habitat for some globally endangered bird species while the cork industry provides income for more than 100,000 people. Cork forests act as a retardant to the increasing number of wildfires caused by climate-change-driven high temperatures and drought: the cork oak's bark is a poor conductor of heat and the thick, insulating layer it provides protects the tree from heat. When cork oaks are replaced with pine trees for paper production, this removes a natural barrier to fire. Pine trees are more flammable, and their needles and bark have a tendency to accumulate; these, along with dead branches, easily ignite and spread fire.

The Rio Frio plantation is also a measure of Amorim's renewed confidence, over 25 years after quality scandals threatened to destroy the cork industry. In the 1990s, concerns emerged over cork taint, a contamination caused by the chemical compound 2,4,6-Trichloroanisole, better known as TCA. Its presence, in minute quantities, is what causes wine to taste 'corked'; it gives an off flavour, like a musty dishrag or wet cardboard. TCA is formed by the chlorine compound Trichlorophenol as a response to a naturally occurring mould in cork, though it isn't clear whether the problem got worse in the 1990s or whether the wine trade had been less alert to it before (there are various theories).

Many wine professionals say that contamination rates in the 1990s reached one in 10 bottles; the cork industry insists its prevalence was much lower, and never greater than five percent. Either way, TCA contamination of cork caused whole wine industries – notably in New Zealand and to a lesser extent Australia – to change over to screw caps. Yet the cork industry's initial response was to downplay the issue. Amorim was founded in 1870; the cork industry's attitude was that it wasn't doing anything different; its production processes were the same as they'd always been.

However, when the next generation took over in the form of new CEO Antonio Amorim, in 2001, the company launched a radical overhaul of its quality control. This entailed the modernization of whole factories, especially at their main production site in Santa Maria de Lamas, just south of Porto. One insider told me that in the late 1990s, some of their facilities still had bare earth floors in places. After massive investment, all that has changed, bringing much more automation and a battery of new cleaning and quality-control kit.

The basic challenge that remains, says Amorim's Antonio Mesquita, is that 'we are trying to offer a natural product with industrial behaviour – guaranteed and predictable'. Since cork is a natural product, no two bark boards are exactly the same.

Not that the raw cork looks any different to the way it did a century ago. It is harvested from May to August – a skilled job, and the best-paid agricultural work in Portugal (workers get about €150 [US $175] a day). Trucks piled high with sections of bark roll into Amorim's Florestal facility, 140km northeast of Lisbon. The bark is first sorted into four grades: the first two, by thickness; the third, of lower quality (about half of the total harvest) for granulating; and fourth, boards with yellow-stain fungus, to be used for non-wine industry applications such as flooring and footwear. Then the bark must be seasoned for six months, in long, high piles on stainless-steel pallets (a measure against TCA) covered with tarpaulins.

Once seasoned, the cork goes through a cooking process where it sits for an hour in water at 98°C: this is another procedure to stop TCA and volatile compounds. It also makes the cork flatter and more workable. Then, in a huge warehouse smelling of wood and bark, workers sort pieces into cages. The cork must be used within three days of the heat treatment, another precaution against TCA – so it either goes north for processing or to another part of the same southern plant. Here, they make thin cork discs from thinner sheets of bark: these are for champagne corks, which are made from several different cork components to better survive the pressure exerted by sparkling wine. Sparkling wine corks form their characteristic mushroom shape only when they are extracted from the bottle. Before insertion, the corks are cylindrical but wider in diameter than still wine corks; they are only partly inserted into the bottle. At Florestal, they make up to 12 million discs for sparkling wine corks every day, all X-rayed to check for hidden defects.

The other product coming off the production line at Florestal is part of Amorim's output of 'technical' corks – the term for stoppers made from agglomerated cork granules rather than whole, punched-out lengths of cork from bark board. These now make up about 80 percent of the wine cork market: they're cheaper – about six Euro cents each, against up to €3 each for the very highest-quality natural corks – and it's easier to clean the cork granules of volatile compounds. At Florestal they do this by steam cleaning. Then machines stick the cork granules together with glue to make corks.

Agglomerated or natural cork is also used in the industry's other main products – the 'T Top' and 'bar top' cork stoppers used for many fortified wines and some spirits. They can be made using natural or agglomerate cork together with a wider disc of wood, PVC, porcelain, metal, glass or other material attached.

The amounts of material involved at Amorim's factories are staggering: outside the plant, mountains of discarded cork board and waste tower above the roadway, while huge piles stretch inside dimly lit warehouses too. But nothing goes to waste. Cork bark rejected for wine corks goes into various applications, from flooring to seals in gaskets, from surfacing in playgrounds to a layer in artificial football pitches, from a shock-absorbing layer between concrete joints to the soles of footwear. And all the clouds of cork dust generated in these plants is collected by suction tubing and bags at every machine to provide biomass for the factories' boilers. Together with solar panels, these supply 70 percent of the plants' energy needs.

Amorim's main plant at Santa Maria de Lamas is higher tech. Here, most of the technical corks are made using 'super-critical' cleaning, where carbon dioxide is injected into a chamber containing the cork granules. When they've been moulded, the technical corks are polished, chamfered, washed and dried, before going through a quality-control process using automatic image control, with each cork scanned one at a time. The atmosphere in the plant is frenetic, with the constant noise from the machines and the beeping of forklifts zipping around with crates of corks.

However, for pricier bottles – anything over about £15 out of Bordeaux, Rioja, Barolo or any of the great European regions – only 'natural' corks will do, punched in a single piece from one strip of bark. It's partly a question of image and partly ongoing uncertainty about how wines will age under technical corks or screw cap compared to cork. The latter is a subject of exhaustive debate in fine wine circles, involving the rate at which minute quantities of oxygen enter a bottle through its stopper. But suffice to say that at best, the jury is still out.

There are many different grades of natural cork, sorted according to the size and quantity of natural flaws in the bark. All must be cleaned to weed out those contaminated with TCA: a harder task for full corks than for agglomerate. Amorim heat natural corks at pressure to volatize any TCA molecules, then re-humidify them.

The punching of a new cork is done by hand for the highest-grade corks, with workers moving with practised swiftness to push corks out at the right point in each strip of bark, for the right thickness and avoiding natural crevices and flaws in the cork. In some instances, small natural flaws can be filled with cork dust and a binding agent to create a more uniform surface – these are known as colmated corks. Lower-grade natural corks are punched automatically by machines, while some are now made by robots, algorithmically improving their aim.

Then the stoppers are washed and dried, before undergoing quality testing using compressed air and X-rays to find internal weaknesses. Amorim also do lab tests, as well as further technical tests for the highest-grade corks. Finally, corks receive a fine external coating which helps lubrication in the corking machine. Previously, coatings were derived from fossil fuel products, but more and more frequently these

– and the binders used to make technical, or agglomerate, corks – are bio-sourced, based on beeswax or plant-derived products.

So, have cork producers fixed the industry's quality-control problem? British wine writer and scientist Jamie Goode says that based on the thousands of wines he has judged at the International Wine Challenge in recent years, 'cork taint is running around one percent of cork-sealed bottles'. Antonio Amorim, the eponymous firm's tall, patrician chairman and CEO, is more forthright. 'I think today cork is perceived as a reliable product. We have the technology capable of giving the performance the winemaker wants.'

The CEO is also keen to make the case for cork's environmental benefits: 'If supermarkets were coherent about sustainability, they wouldn't be using screw caps or plastic. If you use a screw cap, you are adding to emissions; if you use cork, you're reducing them.' Every tonne of cork bark harvested captures 73 tonnes of locked-up carbon dioxide. Meanwhile, aluminium screw caps remain energy-intensive to make or recycle – and most of them aren't.

But the CEO is optimistic that the influence of environment-conscious consumers and retailers will change wine producers' decisions about packaging. For a start, plastic capsules (that protect the cork and neck of the bottle) are now – gradually – starting to disappear. He also thinks that technical corks will make inroads into the screw-cap market, much as they seem to have more or less wiped out the use of plastic corks.

And after wine corks have served their purpose? While natural cork is biodegradable, it is durable and takes years to decompose. There are growing numbers of cork collection and recycling schemes. ReCork operates in the US and Canada: it was started by a Canadian footwear brand in 2008. Recorked UK operates a similar scheme in Britain; all Majestic Wine stores, some branches of the Waitrose supermarket chain and mail-order merchant The Wine Society each accept corks for recycling. A recycled cork is never used to make a new stopper, but it can be ground into granules and used to produce insulation, flooring, sports surfaces, footwear or as a mulch. And carbon remains sequestered internally. Yet because cork is not 'recyclable', it is considered by the UK government to have a high environmental impact and may incur penalties in the new Extended Producer Responsibility legislation, despite its overall sustainable nature. This is surely an example of legislation and taxes not being aligned with a government's stated sustainability goals.

Other kinds of stoppers

Synthetic stoppers became popular in the 1990s as a solution to reduce cork taint. They are a direct substitution for natural cork and aim to replicate cork's compressibility and elasticity. The inside of the plastic stopper has a cellular-foam, spongy structure and they are inserted using the same bottling equipment as cork. They allow a certain amount of oxygen to pass through the stopper from the external atmosphere into the wine. Premium synthetic stoppers are engineered to precisely control oxygen transmission over time, so winemakers can manage the wine's evolution in the bottle. Analysis by PwC showed the carbon footprint for synthetics to be 14.83g per stopper. Synthetic stoppers can be designed to look like cork or made in a range of colours, and can be printed and branded like natural corks.

Synthetic wine stoppers are generally made from low-density polyethylene (LDPE) plastic. Mostly they are produced from virgin fossil-fuel-based plastic, but they can contain a proportion of recycled material. In theory, LDPE can be recycled but actual recycle rates are low and the small size of stoppers makes them difficult to collect and separate. There are no consistent or widespread collection schemes, though some producers have implemented collection points working with wine shops. LDPE, which is not biodegradable, is also used for plastic wrap to stabilize wine cartons on pallets. Synthetic stoppers require a corkscrew, can be more difficult to extract than a cork and are often difficult to reinsert into the bottle. Some stoppers made from plant-based materials such as sugar cane are also now commercially available.

Synthetic stoppers come in a variety of specifications and at a range of prices. Some can cause flavour scalping (loss of some aromatic, fruity aromas) and oxygen transmission rates can be high – up to 20mg per year – meaning that wine shelf life is reduced and relatively short. Cheaper synthetic stoppers can be less forgiving and less elastic than cork. They can be damaged during insertion, leading to problems of oxidation or leakage. Some have been found to allow permeation of external aromas, which can then be absorbed by the wine, leading to instances of 'cork taint'-type characters.

The most widely used alternative to cork stoppers, however, is screw caps, which can be used on both glass and PET bottles. In Europe, in 2021, aluminium screw caps' share of the wine market grew to around a third for light, still wines; screw caps' worldwide share is similar. Consumers and sommeliers value the ease of opening with no need for a corkscrew. A 2020 study in France, Germany, Italy, Spain and the UK showed a preference for aluminium screw caps over cork in the 18–34-year age group. Aluminium screw caps are widely used for spirits (90 percent) and other beverages. As mentioned, their use for wine increased when the incidence of faulty, corked bottles was high in the 1990s–2000s. New Zealand winemakers took concerted action in shifting to aluminium screw caps, which successfully helped to change consumer perception, and today more than 90 percent of New Zealand wines use then. Screw caps come in a wide variety of colours and can be branded as well.

As a manufactured product, screw caps are less variable than natural corks, although if damaged during filling they can lead to leakage and oxidation. The body of the screw cap can be made from aluminium or plastic. The liner is composed of multiple layers of plastic materials with oxygen barriers. Different liners in the screw cap allow differing amounts of oxygen transmission over time: Saran Tin has an extremely low oxygen transmission rate, while that of Saranex liners is higher, although wine faults such as reduction and sulphurous notes can occur.

But aluminium screw caps have the highest carbon footprint of all wine closures, at more than 37g per cap. And while aluminium is infinitely recyclable, the small size of screw caps makes them difficult to detect at materials recovery facilities and most are lost. Collection schemes vary by region, with some recommending that the screw cap is put back on the empty bottle in the recycling bins. For those that are recycled, the liner is usually burned off, with the potential to release toxic chemicals during the process. Both liners incorporate Saran as a constituent part: this is Dow Chemical's registered trade name for polymers of polyvinylidene chloride (PVDC). PVDC is toxic when burned, producing carcinogenic dioxins, and its use is restricted in certain markets, though PVDC-free liners are now available.

Plastic screw caps can be made of polypropylene and polyethylene, though they are much less common than aluminium for wine. The body of the cap contains the same Saran-based liners, and the caps – though not the liners – are fully recyclable.

Another alternative is the 'glass' Vinolok stopper, produced in the Czech Republic by a company now owned by Amorim. While described as a glass stopper, Vinolok is composed of two parts: a glass body and a sealing ring made of Ethylene Vinyl Acetate (EVA) copolymer. It is this plastic ring that creates the seal with the rim of the bottle, so it is more akin to a screw cap than a stopper: glass directly on glass can never create an airtight seal. Vinolok requires a specifically compatible bottle, and the seal seems vulnerable. The cost of Vinolok is high, and they must be applied either manually or on an adapted packing line, adding further to the cost. The EVA ring can be recycled in the same stream as LDPE, but recycling the stopper requires consumers to separate the gasket and glass parts and put them in separate recycling bins.

So far, a lightweight bottle produced from recycled glass and sealed with a natural cork is the lowest-carbon alternative.

Alternatives to glass: plastic and other bottles

At Packamama's manufacturing plant in Suffolk, England, pellets made from flaked PET soft drink and water bottles are shooting through pipes and into machinery to heat them, before they end up in a bewildering variety of stainless-steel moulds and blowing machines. Almost magically, a machine opens and out swings a line of the wine trade's latest plastic packaging solution: a recycled, partially flat, 75cl bottle made of rPET.

'When glass is your benchmark, the material is very important,' says Santiago Navarro, Packamama's CEO and founder. Up to 40 percent of the bottles' surface area is flat, giving up to 50 percent space saving in transit. They weigh just 63g, as against a 420g lightweight glass bottle. They have around half the carbon emissions of a 500g bottle and producing them uses less than 15 percent of the energy taken to make virgin PET; Packamama's European factory also uses renewable solar energy. There is also a version produced using Prevented Ocean Plastic, a project that pays for plastic collection along at-risk coastal sites and sorts and processes it. The scheme provides income for local communities and cleans up the environment at the same time as producing high-quality recycled plastic. And rPET bottles are recyclable.

PET bottles used for soft drinks are not suitable for wine as PET wine bottles require an oxygen barrier to keep the wine fresh. The material used in Packamama includes both an ultraviolet light barrier and an oxygen scavenger, an additive in the plastic that absorbs or blocks oxygen, thereby preventing oxidation of the wine inside the bottle. However, the company will only guarantee fresh wine for 18–24 months, though they are studying an 80g bottle that would give five years' shelf life, thanks to its thicker walls. 'We want to push non-glass packaging to a level that sets a new benchmark,' says Navarro. 'If you want to create a packaging shift, you also need to bring over winemakers at a higher level, because they look at all packaging with the same lens as they use for high-end products.'

Plastics are not an ideal solution. But as Navarro points out, 'this is plastic that's in circulation – we're not encouraging new oil extraction'. He adds: 'We cannot be plastic free with eight billion people on this planet. The right plastics used responsibly are part of a low-emissions future.'

Packamama is now making 1.5 million bottles a year, and The Wine Society have adopted them for several of their best-selling wines, such as the Society's White Burgundy. They are recyclable for several loops, after which they need to be treated with depolymerization, where enzymes are used to break the PET down into its building blocks, from which virgin-quality PET can again be made. Otherwise, they can be downcycled into park benches and the like. But bio-PET, made from plant sugars, can be used to make net-zero PET energy feedstocks.

So far, plastic bottles' adoption by the wine industry has been limited mainly to the airline and travel industry, where weight is an important factor. PET wine bottles can also be traditionally shaped and made of multiple layers, such as a layer of nylon which acts as the oxygen barrier sandwiched between layers of PET. The downsides of PET are a limited shelf life and variable collection and recycling schemes, with the incorporated oxygen barrier making recycling more difficult; the bottles are generally sealed with aluminium screw caps which also present recycling challenges. However, there are certainly occasions when PET may be more beneficial than glass for consumers – for out-of-home consumption at picnics and sporting events, and at festivals where glass is prohibited.

A more surprising bottle alternative is the 'paper' Frugal bottle, made by Frugalpac. The Bordeaux-shaped bottle is marketed as made from 94 percent recycled paperboard, although the actual container in contact with the wine is a composite metallized plastic liner similar to a bag-in-box. The Frugal bottle weighs just 82g and is sealed with a screw cap. The composite inner is not widely recycled, and consumers are required to separate the bottle's constituent parts into separate cardboard and plastic recycling stream bins. Moreover, filling requires investment in specialized equipment and is extremely slow (630 bottles per hour), so filling cost is high.

Innovations in new plastic materials using biomass to replace fossil fuel feedstocks are in development, although none offer a truly commercially viable available alternative. While the aim to replace fossil fuels with renewable sources is the right one, as with biofuels the impact on land and water use for crops needs to be taken into account when scaling up. The aim is to use waste materials and by-products rather than increase demand for crops for human consumption and animal feed. For example, Aimplas in Spain in conjunction with Bodegas Matarromera, has developed a wine bottle from polylactic acid (PLA). PLA is derived from renewable organic sources such as cornstarch or sugar cane. An inner coating of silicon dioxide acts as oxygen barrier. It is biodegradable under industrial composting conditions but can only be recycled if separated from other plastics. As there are no readily accessible recycling systems, this material is limited in its use at this point.

PEF (Polyethylene Furanoate) promises to be potentially even better. This is a bio-based polymer made from fructose sugar derived from plants such as wheat, corn and sugar beet. Ultimately, the aim is to produce PEF from cellulose from non-edible biomass coming from agricultural and forestry waste. It offers eight times stronger protection against oxygen than PET. It is recyclable and can easily be identified and separated from PET using standard near-infrared technology in sorting and recycling facilities. The first commercial plant is under construction in the Netherlands.

Aluminium cans

The packaging industry is the second-biggest consumer of aluminium after the automobile industry, using nearly a quarter of total global production. This is mainly for beer and soft drinks cans: 180 billion of them are made each year globally, the single product that uses the most aluminium. Aluminium has the advantages as a packaging material of being an oxygen barrier, lightproof, lightweight and recyclable.

Cans are portable and easy to chill, hold a single serve and don't smash, making them convenient on the go and for outdoor events and festivals. US wine can sales soared from $2 million to $183 million between 2012 and 2021. Globally, an estimated US $643 million-worth of wine cans were sold in 2024, a figure predicted to increase five-fold by 2034. Several sizes are available, with the most popular, 250ml, delivered in a 12- or 24-pack to retail stores.

However, the SO_2 in wine reacts with aluminium. Hence aluminium wine cans are lined with a food-grade barrier consisting of a thin, plastic polymer coating. But over time, wine's acidity and SO_2 slowly degrade the lining, giving canned wines a shelf life of around nine months. In addition, cans are susceptible to damage and if they leak, they cause corrosion to other cans stacked on the same pallet. Traditionally, coatings were produced from epoxy resins containing Bisphenol A (BPA), a hormone disrupter which can cause health issues, but alternative BPA-free coatings are now available. Canning lines are specialized and demand significant investment and expertise: most canned wines are filled by a third-party packer rather than at wineries. Minimum canning runs can be high, while shelf life, as mentioned, is short.

But the major environmental drawback of cans is that production of primary aluminium is highly energy-intensive. And mining aluminium's raw material, bauxite, is environmentally damaging: it causes land scarring, produces caustic red mud wastes, and releases toxins such as nitrous oxide, sulphur oxide, fluorides, volatile hydrocarbons and other pollutants. The largest mines are in Australia and China, with China the world's largest producer of aluminium. It takes huge amounts of energy to refine bauxite into alumina (aluminium oxide) and then smelt that to produce pure aluminium: 14,000 kWh of energy to produce one tonne of finished metal. According to the International Energy Agency, producing pure aluminium from bauxite ore accounts for up to one percent of all greenhouse gas emissions worldwide.

Conversely, aluminium can be recycled infinitely. Recycling it uses just five percent of the energy needed to produce primary aluminium – though evidently it must be recycled many times to make up for the energy used in mining, refining and smelting it the first time around. Globally, around a third of the aluminium produced annually is from recycled aluminium, a figure that has remained constant for the past two decades.

While aluminium is recycled from many of its uses, 70 percent of cans are recycled globally, with 76 percent of beverage cans recycled in the EU. But recycling rates vary, from Germany achieving 99 percent to Portugal recycling less than a third of its cans. And drinks cans in landfill stay there for 500 years. This makes comparisons with glass or plastic bottles for overall environmental impact tricky. Analysis for the Finnish alcohol monopoly, Alko, suggests that a 33cl aluminium wine can has a lower carbon footprint than PET or glass bottles – but this was based on the can being made from at least 70 percent recycled aluminium, higher than the average percentage. There are cans with a higher recycled content: Novelis, the world's largest producer of recycled aluminium, makes an 'evercan' that is 90 percent recycled.

However, demand for aluminium is high and forecast to grow as a result of its use in lightweight vehicles and solar panel production. If demand exceeds recycled aluminium already in circulation, this will lead to increased production of energy-intensive primary aluminium. So this is not a circular or closed-loop system. Increasing the recycled content of one aluminium product takes it away from another and just leads to the production of more primary aluminium. Thus, growth in canned wine adds to total global demand for aluminium, resulting in higher greenhouse gas emissions globally.

Composite packaging, bag-in-box and stand-up pouches

Aluminium's properties as an oxygen and light barrier mean that it is used in combination with other materials in composite packaging. Plastic-aluminium laminates are made of a layer of aluminium foil sandwiched between plastic layers such as alu-laminate in bag-in-box and Tetra Pak containers. Such lightweight, composite packaging generates lower greenhouse gas emissions during transport.

Bag-in-box (BIB), used to pack volumes of wine from two to 10 litres, have a sizeable share of the market in some countries – almost half of sales in French supermarkets and hypermarkets, and more than half in Sweden's Systembolaget stores. But BIB represented just four percent of global wine exports by volume in 2021 (and only two percent by value).

In BIB, wine is contained in a flexible, airtight bag with a tap inside a cardboard box. The materials used for bags and taps and the quality of the seal determine the wine's shelf life – anywhere between six and 18 months from filling. The inner bag is usually made of several layers of metallized film and plastics which protect the wine from oxygen and ultraviolet light, to varying degrees. The oxygen barrier can be aluminium or ethylene-vinyl alcohol. The plastic tap acts as a one-way valve that prevents oxygen from getting to the wine after opening. The taps are also made of several components composed of different plastic materials – polypropylene, polyethylene and thermoplastic elastomer. The bags can be susceptible to micro-cracking or abrasion, which cause oxidation and damage the wine, leading to waste.

BIB's advantage is that it is lightweight and efficient to ship. It has the lowest carbon footprint of any wine packaging: greenhouse gas emissions per litre are around 7.5 times lower for a three-litre BIB of wine than a 420g glass bottle. Thanks to its rectangular outer carton, BIB is also efficient to stack, taking up 45 percent less space than bottles filled with the same quantity of wine, which gives an energy saving of 20 percent during transport. Packaging is supplied flat to producers, saving on space and logistics costs. BIBs are easy to open and keep wines fresh for three to four weeks after opening, offering the convenience of wine on tap by the glass.

The main disadvantages of BIB are limited shelf life, cost, high minimum production quantities and low recycling rates. For producers, the cost of BIB packaging and filling is more expensive than the equivalent number of 75cl bottles. Filling requires specialist equipment and expertise, for which few wineries are prepared, so it is usually outsourced to a third party. Minimum print-runs for BIB packaging and filling production runs are high. This, combined with the limited shelf life, means that BIB demands careful inventory management: it is suitable only for wines with high rates of sale, to avoid waste. As for recycling, the outer carton is easily recycled, but BIB requires consumers to separate the component parts – bag, tap and box – into different waste streams.

The Nordic countries have the most advanced recycling infrastructure for composite materials, with specific collection and sorting systems. But few other countries have effective schemes and actual recycle rates globally for composite packaging materials are low: most end up in landfill.

Meanwhile BIB have historically been used for lower-priced wines, reinforcing a perception of lower quality among consumers. The exception is Sweden, where, thanks to the Swede's love of its practicality and transportability, and their unsentimental approach to wine, it is the wine packaging of choice. The BIB Wine Company tried to make bigger inroads into the UK market: it was set up in 2023 to offer consumers a greener choice and change perception of BIB wines. The company partnered with Carlos Ludlow-Palafox, who have developed technology at Enval to recycle composite flexible packaging through microwave-induced pyrolysis. Customers ordered BIB for direct delivery from the website and were encouraged to return the packaging after use, after which it was sent to Enval for processing. However, the BIB Wine Company didn't last long, ceasing trading in April 2025.

Smaller-volume wine pouches and Tetra Pak are also made from composite materials, with similar characteristics to BIB. Drink cartons are made from paperboard, using 21 percent polyethylene as a moisture barrier and four percent aluminium to act as a barrier to light, oxygen and aromas. Cartons are generally not made from recycled material but may be recycled to produce new paper products. Recycling of drinks cartons has increased in the EU, with collection rates of more than half, and higher in places.

Cutting cardboard: reducing secondary packaging

Cardboard is widely used to protect goods during distribution. Whatever the primary packaging, most wines are delivered in cardboard boxes – inside which inner dividers may protect individual bottles from being broken during transit. Cardboard is also used in BIB for the outer cartons, as well as for individual bottle gift packs and as trays for various formats.

Paper and cardboard are produced from cellulose from wood fibre: 42 percent of the total global wood harvest is used to make paper, with the biggest proportion going to the manufacture of packaging and cardboard. Forests are a renewable resource providing they are appropriately managed, with FSC certification. Otherwise, using virgin raw materials can lead to deforestation. Recycled cardboard can contain up to 90 percent recycled content without losing durability or strength. Producing recycled cardboard uses half of the energy and electricity and 90 percent less water than cardboard made from virgin material. Yet despite 70 percent of cardboard being recycled, a large amount still goes to landfill – indeed, paper and cardboard account for around a quarter of landfill waste. There, like all biodegradable waste, paper and card emit methane gas as they degrade.

Corrugated cardboard is made of three layers – an inside and outside liner and fluting with a ruffled shape in the middle. The fluting is glued on either side to the flat layers, providing rigidity and stability. The outer liners can either be from virgin – known as 'kraft' – paper or 'test' paper made from recycled paper. Kraft paper is strong and smooth and often forms the external surface of a carton. The recycled test paper is not as strong or as easy to print on but is cheaper and can be used for the inside liner. The glues used are typically starch-based, derived from the roots, tubers and seeds of plants – so eco-friendly. This cardboard comes in different weights, with single or double layers of fluting, known as 'double walled', which can be used for heavier goods.

Using cardboard sustainably to protect wine should save money and reduce waste. A standard 12-bottle carton weighs about 350g, with dividers adding another 130g. Export cartons are thicker and heavier. Wine businesses should seek out cardboard products with the maximum possible recycled content. But the exact choice of cardboard used – and how cartons are printed, varnished, labelled and sealed – all has an environmental impact.

There is also an environmental impact from the plastic tape used on cardboard boxes – which might be relatively small for each box, but adds up. Pierre Nortje, international sales manager at leading South African producer Spier, in Stellenbosch, reports that when they stopped using plastic tape on cartons and switched instead to a water-soluble glue, it took more than 18 tonnes of plastic a year out of their supply chain.

Cases and kegs

Six-bottle and 12 -bottle wooden cases have traditionally been used for premium wines, especially in Bordeaux. A wooden case is more resistant to humidity and damp than cardboard and may be appropriate for wines cellared over long periods of time. They are usually made from pine and are used for presentation packs or as shipping cases. Pine is a renewable resource provided it comes from an appropriately managed forest with FSC certification. However, wooden cases are heavy. A 12-bottle Bordeaux wooden case weighs around 3kg and, when full, this increases to 23kg. The same wine bottled in 450g bottles and presented in a cardboard carton would weigh around 15kg – none of which makes any difference to the wine's drinking quality.

One option that reduces packaging, for bars, restaurants and wine shops that sell by the glass, are draught and keg wine systems. Steel kegs like those used by London company Sustainable Wine Solutions use the same technology as beer kegs, powered by gas, and are long-lived and recyclable. The technically simpler alternative are plastic kegs with a bag of wine inside, under pressure, so that oxygen is kept out and the wine stays fresh. One 20-litre keg holds over 26 bottles of wine, thereby saving around 26kg of glass as well as capsules and cardboard. It reduces the carbon emissions of the wine's transportation too. They're not perfect: plastic kegs like those manufactured by KeyKeg and PolyKeg are single use and not recyclable. Still, keg wine on tap is enjoying an increasing presence, with an increasing range available, including natural wines and even village-level burgundies. Uncharted Wines, one of the biggest suppliers with its Wine on Tap range, now offers more than 60 wines. They claim to have sold almost 12,000 kegs to 220 bars and restaurants over the past year. Such systems are said to keep wines fresh for over two months, which will be enough for most bars and restaurants.

Wine labels

Paul Jones, sales and technical director of Amberley Labels, is proudly showing off one of the company's metallized wine labels: the silver of the design is digitally applied at a thickness of just 0.002 microns and is cured with ultraviolet light rather than using heat. 'EPR [Extended Product Responsibility] and PPWR [the EU's Packaging and Packaging Waste Regulation] has really made people aware of what's needed to be sustainable,' says Jones. He says that most of the papers they use are now recycled. His company also recycles the backing liners from self-adhesive wine bottle labels, as well as stocks of labels made obsolete when a product has changed.

Labels are a small part of wine packaging, but they're vital in providing product information and crucial to a wine's image. They also have an impact on the environment. Labels are used not just on wine bottles but also on the industry's other product packaging. They may be paper-based or plastic; either way, they need to be easily applied, whether using self-adhesive or water-based glues. They need to withstand the wear and tear of the distribution system, and wine bottle labels need to stay on in an ice bucket.

It might not seem obvious, but paper manufacture is an energy- and water-intensive process: turning trees into paper involves huge, energy-hungry machinery to smash up logs. Indeed, worldwide, the pulp and paper industry is the fifth-largest consumer of energy, accounting for four percent of the world's energy use.

Paper mills can also have an impact on their local environment through waste and water use. They traditionally used chlorine-based bleaches, which can result in toxic materials being released into water, air and soil. They also use starch and fillers such as calcium carbonate and titanium dioxide.

Recycled paper needs 70 percent less energy to produce. However, as paper is recycled, its fibres become shorter, meaning that recycled paper requires the addition of around 25 percent virgin fibre for strength and quality. There are now alternative raw materials for paper production, including agri-waste, grape waste, cotton, hemp and bamboo: Dutch company PaperWise makes paper and gift boxes from agricultural waste – stems and leaves of crops such as rice, wheat, barley, cereals, corn, hemp and sugar cane. PaperWise recover the cellulose using 100 percent renewable energy from anaerobic digestion and biomass installation, and use less water in production. It is 100 percent tree-free, 100 percent recyclable and 100 percent compostable. PaperWise says its product's environmental impact is 47 percent lower than FSC paper made from trees, and 29 percent lower than recycled paper, with these claims certified by the University of Amsterdam. Labels are now also available made from calcium carbonate, without any paper content at all.

Paper labels also involve printing, with a range of inks, films and lamination treatments to create the desired effect. Conventional inks use mineral-based oils, with pigments containing heavy metals such as mercury, volatile organic compounds, polychlorinated biphenyl and chlorofluorocarbons. These do not biodegrade and can cause pollution and contaminate recycling processes. Bio-renewable ink is derived from tree, plant, insect and/or animal materials, including resins, gums, oils, waxes, solvents and other polymer building blocks, but not all are biodegradable. Biodegradable inks use natural vegetable oils, such as soya, sunflower or rapeseed oil.

For recycling, conventional paper wine labels disintegrate in the sorting and preparation of glass cullet. Printing inks, foils, laminates and adhesives can generate emissions as they burn in the glass furnace. Plastic labels, more firmly bonded to the glass, are more difficult to separate: pieces of glass that remain stuck to the label can affect smelting.

Meanwhile, for PET bottles, the labels must detach easily from the bottle in the grinding process and float to the top of the water basin into which PET bottles go for separation. Printing inks, foils, laminates and adhesives can likewise contaminate recycled PET. Research has suggested ways around some of these problems: using label substrate compatible with PET recycling together with soluble inks and glues would improve the overall recyclability of PET bottles. Self-adhesive labels are delivered on rolls of backing material made from coated paper or plastic film with a silicone coating and are difficult to recycle.

Capsules

Capsules are traditionally placed over the top of a stoppered wine bottle. Once upon a time, they helped protect corks, but nowadays they are there mainly for aesthetic and brand-identity purposes. Capsules are made from tin, aluminium, a complex polylaminate plastic, or PVC. While a wine drinker cuts off the top of the capsule before taking the cork out, the main body of it usually stays on the bottle neck, which has an impact on recycling.

Polylaminate capsules are a blend of aluminium and polyethylene. PVC can in theory be recycled, but the US Environmental Protection Agency estimates less than three percent actually is. PVC contains toxic phthalates and heavy metals and creates dioxins when burned. Meanwhile, sparkling wines generally use a longer capsule foil over a steel wire cage and a metal cap known as a muselet to secure the cork. The only capsule material that is recyclable in practice is tin, which is energy-intensive to produce, though, like aluminium, can be infinitely recycled. However, according to the International Tin Association, the recycling rate for the metal in 2019 was less than third.

The best sustainable capsule alternative for producers nowadays is simply to dispense with them altogether. The UK's Wine Society reports that getting rid of the capsule on the Society's Claret, one of its bestselling bottles, saved around 435kg of plastic. Altogether, in 2024, the Society sold 320,000 bottles of wine without capsules. Some producers have done the same, with no apparent impact on sales.

Conclusion: going green with packaging

Greener packaging is complicated, as one wine brand's choices show. In June 2022, Accolade Wines launched a sub-brand of its widely popular Australian brand Banrock Station – Wise Wolf – using packaging made with 95 percent recycled materials. The glass bottle, produced from 100 percent recycled glass, has to be heavier and is sealed with a screw cap. To help compensate for the weight, the bottle neck has been made shorter to use less glass. This means less cardboard is used to box the wine and allows more efficient stacking on transport pallets, with 20 percent more wine on each. The wine is packed for Accolade in the destination market at Encirc in Bristol, in the UK, which uses 100 percent renewable electricity and has a zero-waste-to-landfill policy. The site has been awarded UK Sustainable Manufacturer of the Year for several years. Accolade acknowledge that they are on a journey, and that all is not perfect. In particular, the Saranex screw cap liner is identified as an issue, as well as silicon dioxide necessary in the recycled glass.

It's just one example of the multiple decisions and challenges involved in the packaging of a single product. For a producer or retailer to go green on wine packaging is complicated because they need to consider each component, both in isolation and as part of a whole sequence. Every choice has implications for both greenhouse gas emissions and waste.

Nevertheless, the most important general principles are to reduce packaging weight and to use packaging that is easily recyclable, and which is made from recycled materials. And for the wine trade in importing countries, understanding what packaging manufacturers are doing to reduce their Scope 1 and 2 emissions will help the trade reduce its Scope 3 emissions (*see* page 48). This means asking questions about packaging suppliers' sustainability credentials, making careful choices about the best companies to work with, and being attentive to all possible packaging options.

However, for wines intended to be drunk more than 18 months after bottling, glass remains the only viable option for the moment. The lightest-weight option, with a high recycled content, will be the most carbon-friendly container. Switching to lighter-weight bottles means an immediate reduction in greenhouse gas emissions – and is easy to implement. Whether the bottle is sealed with an aluminium screw cap or cork, with or without a capsule, and packed in a cardboard carton or wooden box can make a material difference to emissions and the wine's place in the circular economy too. Wines made for earlier drinking have a wider number of options – but with trade-offs to bear in mind.

Key lessons for sustainable packaging

- **Reduce glass bottle weight** and increase recycled content.
- **Minimize packaging weight** – producers should consider primary, secondary and tertiary packing holistically and in the context of the supply chain. Seek out maximum recycled content for cardboard.
- **Move towards a circular economy** that reduces waste, reuses and recycles packaging.
- **Consider alternative packaging formats.** Businesses in the on-trade should consider larger volumes such as kegs and reusable formats.
- Use **corks as a preference over screw caps.** Consider getting rid of unnecessary packaging such as capsules.
- **Eliminate use of plastic** films and tape.
- **Source packaging as locally as possible** – for example, bottles, cardboard.
- **Understand recycling schemes in destination markets** and make changes to your packaging accordingly.
- **Exert pressure on suppliers** to reduce bottle weights and offer a wider range of packaging options; lobby local governments for effective collection and recycling schemes.

Chapter **FIVE**

WINE'S SUPPLY CHAIN

In the car park, the hum of three huge wind turbines' rotor blades announces the size of Lanchester Wines' Greencroft bottling plant, near Durham, northeast England. Inside the sprawling plant's vast buildings, there are nine production lines: six for bottles, two for bag-in-box wine, and one canning line. At one end of the plant, bulk wines arrive in 24,000-litre flexitanks in shipping containers; at the other, cartons of filled bottles are stacked high in the warehouse – there are 40–50 million bottles under their roof at any one time. In the bottling plant, there aren't a lot of workers: most of them appear to be either driving forklift trucks or fixing machines. The huge production line is almost entirely automated, from labelling machines to palletizers.

Today, bulk-shipped, UK bottled wine accounts for at least 40 percent of wine sold in the UK, including 18 of the top 20 wine brands. That includes independents as well as supermarkets, plus a big chunk of the off-trade market (Greencroft do 18-litre bag-in-box wines for one major pub chain). But this is just one of wine's routes to market, and it raises a series of complicated questions about sustainability.

The shape of the global wine trade

Italian wine company The Wine People produces wines from Sicily, the Veneto, Abruzzo and elsewhere and has a strong sustainability record. Their production has been organic for years: 'Sustainability was a natural progression,' says owner Stefano Girelli. Nevertheless, getting the 9.5 million bottles it exports annually to consumers is a complex operation. It exports to 36 countries: the UK is its biggest market, taking 38.5 percent of its exports, followed by the Netherlands (23.8 percent), the USA (5.6 percent), Belgium (3.2 percent), Germany

(3.2 percent), Japan (2.7 percent), and then a long tail of smaller totals for the remaining 30 markets.

As with all wine, The Wine People's gets from the producer into consumers' glasses via a complicated global supply chain. Indeed, wine logistics and distribution account for 10–20 percent of wine's total carbon footprint, as well as representing a significant share of the price tag of any bottle. Of course, wine is just one product in a web of international logistics. International Transport Forum figures show that trade-related freight transport generates around eight percent of total global greenhouse gas emissions, equivalent to three billion tonnes of CO_2. Creating a more sustainable supply chain means minimizing carbon emissions from transport and distribution, and creating a circular economy.

All major wine regions and big wine companies depend to some extent on exports. One in every two bottles of wine produced is sold cross border: many of these are transported from one side of the world to the other. Grands Chais de France, one of France's largest wine companies, exports to over 178 countries, accounting for 80 percent of its turnover. In 2021, Chile exported 83 percent of its wine production; New Zealand exports 88 percent of its wine and Australia 71 percent.

On the import side, in 2024 the US remained the world's largest consumer and importer of wine by value (in addition to being the fourth-largest producer). Over half of US wine imports come from Europe. The largest importer by volume, Germany, brings in wines from neighbouring countries Italy, France and Austria as well as from Spain and as far away as South Africa. The UK imports large volumes from Australia, Chile, the US, South Africa and New Zealand, as well as from Europe.

To reach different international markets, some wine producers deal directly with retail companies, while others appoint agents in the importing country to represent them. In turn, agents may sell on to other agents and trade customers – as US importers are forced to, under the oddities of their country's three-tier system introduced in the 1930s after Prohibition. In larger export markets, some producers may have an office or shared ownership of a sales and marketing company which represents them. There is no one dominant way of being present in a market; the result is a complicated and fragmented network of trade intermediaries.

One of the main characteristics of the wine trade is its fragmentation at all levels: this has a big impact on its sustainability. A vast number of wines are produced by many different companies. It is estimated that there are over a million wine producers worldwide: the world's largest, California's Gallo, makes just three percent of global wine production, while the top 10 wine producers collectively make around 12 percent of the global total. By contrast, brewing giant AB InBev produces more than 25 percent of the world's beer, and together with Heineken and Carlsberg accounts for around half of global production. The top 40 brewers make 86 percent of the world's beer. Not only is the structure of the wine industry different to that of beer and spirits, the most valuable wine brands, ranked by consultants Brand Finance, are worth a fraction of the leading spirits and beer brands. Champagne brands Moët & Chandon and Veuve Clicquot, ranked one and two, are valued at $1.4 billion and $962 million respectively – compared to Corona beer at $7 billion and number one spirits brand, China's Kweichow Moutai, at $42.9 billion. Wine companies are smaller, with fewer resources.

At the consumer end, the wine trade is split between two main channels. What is known in the UK as the 'on-trade' refers to hospitality businesses that sell wine for consumption on their premises – bars and pubs, restaurants and hotels. The 'off-trade', where 80 percent of wines are sold globally, is dominated by grocery supermarket chains, alongside specialist drinks chains and independent wine shops. But it varies a lot by country. In the US, over 75 percent of wine sales are off-trade, against 80 percent in the UK. The proportion of wines sold through the on-trade also varies by market and is often serviced by specialist agents and distributors. Thus, in France, the big supermarket chains represent around half of wine sales, while three in 10 bottles are sold in restaurants, and one in 10 from direct sales from the winery, in person or online. German wine sales, by contrast, are dominated by big discounters, who make more than 60 percent of total wine sales, while only a relatively small proportion is sold in the hospitality trade. Wine-selling regulations and licensing laws also vary a lot from country to country.

Another key difference between wine retailers and others is the high number of different wines that they sell compared to other product categories. Grocery stores often stock 750-plus wines nationally, while specialist shops can sell thousands: London's Hedonism stocks 3,500 wines, Lavinia in Paris 6,500, and US specialist drinks retailers Total Wine & More, in California, carries 8,000 different wines. Ranges constantly evolve so that previously lesser-known regions become mainstream in shops and on restaurant wine lists – such as, over the past decade, Austrian Grüner Veltliner, Oregon Pinot Noir, Elgin in South Africa, the Limarí Valley in Chile, and English sparkling wines. But in practical terms, the high number of product lines in a range means low average rates of sale: few wines sell in large volumes and a large number sell in small quantities.

Selling wine to consumers

Penny Edwards is very proud of her wine shop, Cellar Door Wines, which sits on a small industrial estate just outside the centre of St Albans, Hertfordshire, in the heart of London's commuter belt. Originally hailing from Botswana, she is a veteran of the UK wine retail trade, having started out at Oddbins in London in her early twenties. Cellar Door Wines lists 700–800 wines at any one time and specializes in South African and Georgian wines, among others. But the shop that she manages and co-owns has a turnover of just £700,000 (US $940,000) a year. She cites the pressure of rising costs in a time of high inflation; the doubling of business rates (the UK's local business taxes); 2025 increases in both employers' National Insurance (payroll taxes) and the minimum wage; as well as a series of alcohol duty (tax) hikes imposed by both Conservative and Labour national governments. Her firm is too small to be charged the new Extended Producer Responsibility (EPR) fees for packaging waste, starting in October 2025, but she says: 'EPR costs are passed on to us both via suppliers and our landlord in terms of waste removal from our premises.' Cash flow is a constant worry: 'We always have a lot of stock of interesting products, but never enough of the one that the customer wants!', she says.

Small and medium-sized retail companies and hospitality businesses like Edwards's account for a significant proportion of global wine sales. Specialist wine shops like hers may specialize in wines from a particular region or a tailored offering, such as organic or natural wines. The strength of specialist companies is their expertise and depth of understanding of wine as well as their direct relationships with customers. These companies are an interface between wine producers and consumers: hand-selling wines and providing advice and a high level of service. Often these companies are owned and run by passionate wine experts – like Edwards – and many have trade qualifications such as the Wine & Spirit Education Trust (WSET) awards or diploma. Talking to Majestic Wine customers in the UK, the strong personal relationship between customer and an individual in their local store is equally, if not more, important than the range of wines on offer.

But smaller wine shops are under threat due to economic pressure: Edwards cites competition from grocery chains and online merchants, while she offers a service which is more costly to deliver. The improved range and quality of wines on offer in the supermarkets, combined with the convenience of buying wine at the same time as the weekly food shop, have changed customer shopping habits. In the UK, Majestic Wine, which accounts for five percent of off-trade wine sales, is now the only remaining national specialist drinks retailer. Previous big-name off-trade chains Victoria Wine, Peter Dominic, Thresher, Wine Rack and Oddbins have all gone bankrupt or merged over the past 20 years. Nevertheless, independent wine retailers with just one or two shops have seen something of a resurgence in the UK in recent years, with 1,018 shops run by 774 companies in 2023, up by half in around a decade. There has also been a rise in the number of trained sommeliers in restaurants in the UK.

Many bricks-and-mortar wine retailers – including Cellar Door and Majestic Wine – also offer wines online; other wine companies have a purely online presence, including Naked Wines in the UK, US and Australia, and Wine.com and Vinfolio.com in the US. Online sales increased during the COVID lockdowns. Direct-to-consumer (DTC) sales at the cellar door are important in many wine-producing regions, while online DTC sales are important in some regions such as California. Producers in some regions have a high proportion of direct sales, having developed wine tourism and clubs. Online sales have been aided by the rise of mobile technology. Customers can now easily search for wines, compare prices, find vintage and product information, tasting assessments, vintage reports, and check availability. There is also an on-going shift in use from pure information to video and sharing experiences on platforms such as Instagram and TikTok. Meanwhile, online virtual tastings, which, again, mushroomed rapidly during the COVID lockdowns, have become a permanent feature and can bring together groups of people around the world.

Fine wine is something of a special case in wine retailing. It tends to be traded through specialist channels. 'La Place de Bordeaux' operates as a clearing house for fine-wine sales: established initially for the classed growths of Bordeaux, it has expanded to include other premium wines. Fine-wine buyers are buying wider portfolios from selected estates in Champagne, Tuscany, Piemonte, the Napa Valley, the Barossa Valley and the Rhône. Individual estates in South Africa, New Zealand, Chile and Argentina are also generating a following for their super-premium wines. Some of these are sold as futures or 'En Primeur' by specialist merchants: this means they are sold while still in barrel and will be matured, bottled and delivered in the future, generally 18 months to two years later. Contrary to popular perception, fine wines, defined by wine think tank Areni Global as priced $75 a bottle or higher, are purchased by a surprisingly young cohort, not all of whom are wine collectors.

Lastly, a key feature of the international wine trade is the huge share of wine sold by large, multiple grocery chains. In the US, more than half of wine sales are through supermarkets, with Walmart by far and away the largest wine retailer, followed by Amazon. Costco (third in the US) claims the title of largest wine retailer in the world, with annual wine sales worth $2.5 billion. In the UK, two-thirds of retail wine sales are through eight supermarkets. And where wine is sold by government-owned monopolies, as in Canada and Sweden, these are also huge organizations.

Since multiple retailers have the largest number of consumers, they quickly pick up on and respond to consumer trends. Through their choice of offering, they shift attitudes and set expectations, for example on price. Ultimately, they can influence behaviour too. The growth in organic food sales, for example, is as much due to wider distribution in supermarkets as to increased consumer perceptions of its health and environmental benefits.

Wine's global supply chains: container ships

To give an idea of one wine's journey, take a bottle of Marlborough Sauvignon Blanc on the shelf at a store in the eastern English city of Norwich. After the retailer places an order through a freight forwarder, a container housing a flexitank arrives at the winery by truck, where it is filled. The container is then driven to the nearby port of Nelson, where it is transferred to a feeder container vessel ship to make the journey up to New Zealand's main international port of Tauranga, on North Island. Here, the container will be offloaded for transfer on to a larger ship. The fastest route from Tauranga to the UK is through the Panama Canal, which takes around 50 days on the water. Depending on the shipping line, it may go direct to London Gateway, a few miles east of the city, or to Rotterdam, where again it will be offloaded and put on another feeder vessel to the UK, adding another seven days to the journey. Upon arrival at London Gateway, the container is offloaded and has to complete customs clearance, which usually takes two to three days. It is then transferred by road to a bottling facility, such as Encirc or Greencroft, several hundred miles away. There, it goes into bottles, which are boxed, palletized and transported by road to a retail distribution centre and subsequently out to retail shops, including those in Norwich.

All in all, the journey takes 70–77 days from winery to shop shelf. However, not all ships are direct, and ships get rerouted, often stopping in Asia; or instead of the Panama Canal, they may go around the Horn of Africa. Other wines follow routes from San Francisco's Oakland port, and from San Antonio in Chile, which is also the nearest departure point for Argentinian wines from Mendoza. But getting to San Antonio requires crossing the Andes by road: the pass is at 3,175 metres above sea level and is sometimes closed by heavy snowfall in winter – in 2024 it was shut for two months. When the pass is closed, wines generally have to cross the country by road to Buenos Aires, some 1,000km to the east.

Around 90 percent of the world's traded goods are transported by sea freight. The journey from producer to consumer usually involves multiple legs over weeks or even months, with numerous intermediaries, freight forwarders, transporters, importers and agents responsible for different parts of the network. Even wines sold in their country of production often travel long distances. Transport and logistics companies often subcontract parts of the journey to third parties. In addition to the physical transport of wine, the whole process can also be delayed by the completion of customs documentation. 'Global logistics into the UK is a major problem, especially since Brexit,' says one senior figure in the UK bulk trade. 'If there are delays in port, we have to pay "demurrage" charges and detention charges. The typical time to turn stuff around from the port is 11 days – but sometimes it takes three days even for us to be notified that it's there.'

Wine is usually transported in standardized 20-foot (TEU) or 40-foot (2TEU) shipping containers (6.1m and 12.2m, respectively) that have come to dominate freight since the 1960s. There are around 5,800 container ships circling the world: on average they carry around 15,000 containers each. Most ships are owned by a small group of multinational companies: MSC and Maersk account for more than a third of the market. The ships operate on fixed routes to predetermined ports. Containers can be transported by road to port, transferred by crane onto a container ship, and off onto rail or road at the destination port for onward transport. Wine's journey invariably involves this 'trans-shipment', the process of transferring goods from one vessel or mode of transport to another to complete a journey.

Generally, shipping by full container load is most efficient for cost and emissions: the cost of a full container load is fixed, so the more wine there is in a container, the lower the transport cost per litre. A 20-foot container's unloaded weight is between 1.8 and 2.2 tonnes and it can be loaded to a maximum total weight of 25 tonnes. Containers are often loaded with wine on wooden pallets – typically stacked four layers high with 56 cases per pallet. Eleven Euro-pallets (0.8m x 1.2m) or 10 standard pallets (1m x 1.2m) can fit in a 20-foot container: that's approximately 7,000 bottles, depending on their weight and dimensions. In some cases, wooden pallets are replaced with 'slip sheets', a less expensive, thin plastic or fibre sheet, which allows more space for goods. Wine boxes can also be loaded directly by hand into a container to maximize container space, but this requires a lot of labour.

Containers are used multiple times, circulating around the world with an average lifespan of around 20 years. They are built to withstand extreme weather conditions, as they may be stacked on the deck of a container ship: they can be completely sealed and are wind and watertight, providing security and pest prevention. But the temperature inside a container can vary substantially: a container sitting on a dock or by the roadside in a truck in the sun can heat up to more than 40°C. Insulated containers have an insulating liner that reflects heat and reduces excessive temperature spikes. Meanwhile, refrigerated containers – known as reefers – are fitted with a cooling unit to maintain an optimum temperature and protect wine quality. They are smaller in capacity due to the space taken up by the refrigeration unit. They generate nearly a quarter more greenhouse gas emissions due to the energy used for cooling and their smaller capacity.

Wine retail and hospitality businesses generally require a flow of wines with just the right amount delivered frequently to keep them in stock and maximize sales, without taking up space and having large amounts of money tied up in inventory. The large range of wines on offer, many of which sell in only small quantities, creates a long tail which makes managing inventory challenging. Businesses need a high level of coordination to ensure that goods are delivered in good condition, on time, and to meet trade customers' expectations. So stock generally has to be held somewhere upstream to enable flow into stores, in a holding depot owned by a third party. For delivery to store or direct to customers, some retail companies have their own vehicles, or they may contract a delivery service; many utilize a combination of both.

Wine does not require the same strict storage and handling conditions as some food products, but conditions are important for preserving quality and preventing damage. Wine can be degraded by temperature, light and humidity. Ideally, wines should be kept at a constant temperature, between 12 and 14°C, with relative humidity at 55–80 percent, though they can withstand short periods at higher temperatures provided these are not too extreme. Wine stored at over 25°C for an extended period or exposed to 40°C+ even for a short time may change irreversibly – and, as mentioned, wines sitting in a container on a dock or roadside in a truck in the sun can easily suffer in this way. Heat can cause oxidation, premature ageing, a reduction in fruity aromas and colour changes. Heat can also cause corks to lift, bottles to leak and swelling in bag-in-boxes. Humidity can cause mould growth and damage labels.

Conventional supply-chain management reduces costs, optimizes inventory and ensures regulatory compliance. Sustainable supply chain management additionally includes the goals of reducing carbon footprint and encouraging a circular economy. So understanding where emissions occur at each stage of the route from producer to consumer is important. Planning for what happens to packaging materials across the supply chain is essential too, if businesses want to move towards a circular economy. Businesses must also manage climate change-driven disruption of their logistics.

First, note that the weight of transported goods correlates with their greenhouse gas emissions: a heavy product like bottled wine means more CO_2 emissions, whatever the mode of transport. As examined in Chapter 4, 85 percent of wines are sold in glass bottles. For wines that will be drunk 18 months after filling, glass is the only suitable material to maintain product quality – but it is heavy compared to other packaging. Since bottle weights vary from under 400g to over 1kg, transport emissions can double based solely on a producer's choice of glassware. Lighter-weight bottles and alternative forms of packaging will reduce a producer's Scope 3 transport emissions.

Second, just like sending a parcel through the mail, the cost also depends on how much space the wine takes up. Using space efficiently reduces emissions per unit of product as well as the cost. We saw in the previous chapter how wine packaging can have a radical impact on the space the bottles take up – or how many of them can be fitted on pallet. Thus, flat rPET bottles and bag-in-boxes use space more efficiently.

Bulk wine shipping

Outside Encirc's Cheshire bottling plant, there is a line of truck trailers loaded with 20-foot shipping containers. Inside each container is a flexitank, essentially a 24,000-litre, single-use composite plastic bag. These were filled weeks ago at wineries in Australia, Chile and South Africa. Workers have connected hoses to one container, from which they are pumping the contents into 26,000-litre stainless-steel tanks in the vat room next door. Deeper inside the plant, bottles are being filled, labelled and put in cartons on a bottling line that can handle 28,000 bottles an hour. Encirc make all these bottles themselves, some of the two billion glass containers a year emerging from their glass furnaces, which sit next door to the bottling plant. Bulk wine's quality has improved dramatically, especially in terms of oxygen control. 'From a quality perspective, we'd argue that [shipping in bulk] is better for quality than shipping in bottles,' says Encirc's Robin Thompson. 'The temperature profile [during transport] is lower in tanks – it's better than if the wine is sitting on a dockside in Singapore in bottles at 40°C. We rarely see issues now. We're pushing the envelope and doing the next tier up now, French Pinot Noir, Marlborough Sauvignon Blanc.'

Certainly, the UK bulk shipping and bottling trade is in healthy shape at present. Encirc bottle more than 330 million litres of wine a year at their plants in Cheshire and Bristol. Greencroft pack 180 million litres a year, in bottles, bag-in-box and cans. In total, at least 40 percent and perhaps 50 percent of wine sold in the UK is shipped in this way; one senior industry figure thinks that it can go higher, to perhaps 700 million litres a year – which would be more than 60 percent of sales. It is a change from the days when British merchants imported barrels of Bordeaux and port to be bottled in the UK. Even Bordeaux First Growths – in the late 19th up until the mid-20th century – travelled before bottling, also by sea. That trade almost disappeared. Ironically, one Greencroft staff member says he left wholesaler Waverley's bottling plant when owner Scottish & Newcastle closed it in 2007, telling staff there was no future for bottling in the UK.

Greencroft's operation, with a fully automated tank room and a canning line that can fill 17,000 cans per hour, in addition to their bottling lines, is impressively sustainable too. As well as their three wind turbines, they also have a three-million-kilowatt solar array on the roof, which more than meets their electricity needs – so they export to the grid. Buildings are heated via a heat exchange system using water from a nearby disused mine – though their heating needs are 'negligible', thanks to a recyclable cladding system with as much insulation in the walls as the roof. They recycle flexitanks, broken bottles, cartons and plastic shrouds from pallets – though they are not yet fully zero waste to landfill.

Bulk shipping is attractive to importers and exporters primarily because it is cheaper than shipping bottles. But since the wine itself accounts for most of the weight being shipped, as well as lower shipping costs per litre, bulk shipping has the lowest carbon footprint per litre. So while Britain is the largest export market for Australian wine by volume, representing a third of Australian wine exports, 89 percent of that is shipped in bulk and packed in the UK, including Hardys, the

UK's number-one-selling wine brand. To transport the same amount of wine in bottles on pallets would require an additional three shipping containers. Because 35–40 percent of the greenhouse gas emissions caused by shipping are due to the weight of glass, the bottled wine has a higher 'emission intensity' per litre.

Of course, the bulk wine must still be packed at its destination before it can be sold to customers, adding cost and generating additional greenhouse gas emissions which need to be counted. But if the wine arrives at a port close to a modern, efficient packaging plant such as Greencroft or Encirc, overall emissions should be lower than if it had been shipped in bottles. Analysis done by Carbon Intelligence for Encirc's Spanish owner, Vidrala, indicated that bulk shipping and bottling at destination can reduce shipping carbon footprints by more than half. Bottling at destination also helps even out the mismatch between where goods are produced and consumed, helping recycling by closing the loop.

Shipping 24,000 litres of wine is only feasible for wines sold in volume with a high rate of sale. In other words, in practice it is used for widely distributed brands and wines for large multiple grocers. Wine can be bulk shipped in smaller volumes but the impact on carbon footprint in this situation is less clear: ISO tanks are reusable stainless-steel tankers, which can be subdivided into two or three smaller compartments. ISO tanks are commonly used to transport wine between wineries and bottling plants in wine regions or further afield. They can be cleaned and reused, but they are heavier and hence generate higher emissions than a lightweight flexitank. Smaller bulk quantities, around 1,000 litres, can be shipped in intermediate bulk containers (IBCs) or drums. There may be some savings in emissions from the transport from producer to filling line, but a bottling line working on small volumes is likely to be less efficient: it is not at all clear that such a situation would generate a lower overall carbon footprint per litre of wine.

But bulk shipping also raises tougher questions: if packing shifts away from the wine production region, the producer loses a share of the profit margin – and there is a potential impact on jobs, too, often in rural areas. In the introduction to this book, we heard RJ Botha, winemaker at Kleine Zalze in South Africa's Stellenbosch, with his pithy view: 'I have a big issue with shifting bottling from a country with 37 percent employment to a relatively rich county. We need to create as many jobs as possible here.' Another South African, Carolyn Martin of Creation, in Elgin, is similarly critical: 'I'm not in favour of South African wines being shipped in bulk to be bottled elsewhere.' There is a conflict here, in other words, between environmental and social sustainability, between reducing greenhouse gas emissions and reduced profit and employment.

A business shipping less than a full container-load of cases of packaged wine can use 'groupage'. Several smaller loads from different companies are consolidated and packed into a container. It is generally more expensive and takes more time, as the separate loads of wine must be combined to complete a container. In this instance, goods are generally transported from the wine producer to a warehouse by truck, where they will be stored until the rest of the container goods are received. It is then all loaded into a shared container, transported by road to the port or rail terminal, and customs documentation, duties and taxes approved prior to loading. On arrival at the destination terminal and after customs clearance, the load is transported by road to a warehouse, unloaded and separated, before being transported onwards to the importer.

Cleaner transport

Transport remains heavily reliant on fossil fuels: the sector accounts for 30 percent of global energy consumption via rail, air, road and sea, these being energy-dense oil products (petrol and diesel) which generate a disproportionate share of greenhouse gas emissions (37 percent of the total) as well as air pollution. Overall, International Transport Forum figures show that trade-related freight transport generates around eight percent of total global emissions each year, equivalent to three billion tons of CO_2. Developments in electric vehicles (EVs), battery and hydrogen technology will help electrify transport systems – but, on the whole, shifting to low-carbon fuels in the transport sector is not easy. In particular, the areas where change is toughest are aviation, maritime shipping and heavy road freight.

Shipping is the most efficient mode of transport according to the standard measurement of carbon dioxide equivalent emitted per tonne-kilometre, ie, transporting a tonne of goods over one kilometre (abbreviated as CO_2 e/t/km). Moving goods by sea generates 7gCO_2 e/t/km by sea – much less polluting than road (137gCO_2 e/t/km) and a fraction of air transport (654gCO_2 e/t/km).

Nevertheless, the sheer volume of shipping in today's trade system means that its total emissions are still larger – around three percent of global emissions – than those of aviation. Almost a quarter of those emissions are due to container shipping of the kind relied upon by the wine trade. In addition, traditional ship fuels, such as heavy fuel oil, are among the dirtiest, with particularly heavy emissions of sulphur dioxide. The International Maritime Organization rates the energy efficiency of ships from A to E as part of measures aimed at cutting the shipping industry's carbon footprint by at least 40 percent by 2030. At present, just one in eight ships is rated A. In the shorter term, there are lower-sulphur fuels available, or ships can be fitted with exhaust gas-cleaning systems ('scrubbers') to scrub the excess sulphur from the exhaust before it is released into the atmosphere. In the longer term, the Global Maritime Forum's Getting to Zero Coalition aims to get commercially viable deep-sea vessels powered by zero-emission fuels into operation by 2030. And the European Commission's 'Fit for 55' legislation sets out a 55 percent reduction in net greenhouse gas emissions by 2030 on 1990 levels.

To achieve this aim, the shipping industry has been investigating alternatives such as biofuels, electric ships and even a revival of sails: in 2024, agribusiness Cargill's container ship *Pyxis Ocean* completed a successful six-month trial using rigid, 37.5-metre sails. And in September 2024, champagne producer GH Mumm sent its first shipment from Le Havre to New York under sail, taking 14 days

– compared to an average of around nine days on a conventional container ship. Mumm's partner, TransOceanic Wind Transport, hopes to expand its fleet; the *Anemos*, the sailing ship that carried the first Mumm cargo, can transport 1,000 tonnes of cargo. It saves more than 90 percent of the carbon emissions of a normal cargo ship. Mumm's aim is to use sail as its main route to the US market. But to put this into context, a reminder that the average container ship carries 15,000 containers, each one loaded up to approximately 22 tonnes, giving a total cargo weight of 330,000 tonnes – all of which dwarfs the tonnage of the *Anemos*.

But it is road transport that generates 73 percent of total transport greenhouse gas emissions, with an average of 137gCO_2 e/t/km compared to rail's 24gCO_2 e/t/km. This is why, while three-quarters of freight travels by sea, it accounts for just 12 percent of freight transport emissions. Many wines are transported across Europe and North America by road. And for short distances, for collection and for delivery to its final destination, road transport is usually hard to avoid for the wine industry or indeed any consumer goods transport. However, even short distances can account for a relatively large proportion of total transport emissions. For example, a container transported 300km in a truck from Hamburg to Berlin generates a third of the emissions (0.3 tonnes) as the same load going by ship from Hamburg to Shanghai, a distance of 20,000km. Consolidation of shipments to maximize vehicle-loading efficiency can lead to longer distances being travelled than the most direct route, so increasing emissions.

Decarbonizing road transport is now a major focus of sustainability policy and of efforts against climate change in many developed countries. At the same time, air pollution and congestion from road transport have driven an increasing number of cities to introduce low-emission zones as a public health measure. For example, London's Low Emission Zone, operating since 2008, means that heavy goods vehicles over 3.5 tonnes must pay from £100 to £300 a day to drive anywhere in Greater London, depending on the Euro rating of their engines. And we have seen significant progress on electric cars: countries including

China, France, Germany, Italy and Japan have pledged to phase out cars with internal combustion engines. Take-up of EVs has grown steadily: the International Energy Agency predicts that more than a quarter of new car sales worldwide in 2025 will be electric. That figure was nearly 30 percent in China in 2024 and just under 20 percent in the UK. And the charging infrastructure is expanding fast in many countries.

Much trickier to decarbonize is the road freight sector, above all long-haul trucking. The sheer amount of power needed to propel a 44-tonne truck travelling long distances means that there is still no viable electric technology for this type of vehicle. There are now some lighter electric trucks on the market, designed to travel shorter distances, such as some electric builders' merchants' trucks. Amazon introduced five 37-tonne, fully electric Heavy Goods Vehicles (HGVs) in the UK in 2022. But the batteries for these vehicles currently weigh around a tonne each: the number that would be required to power a heavy, long-distance vehicle is impractical, though major leaps in battery technology could change that in the future.

One solution being developed in China by battery maker CATL, the world's largest, is a network of stations where standardized truck batteries can be swapped out for charging. Taking out the battery and putting in a charged one, says Robin Zeng, CATL's CEO and founder, will take less time than filling a truck's tank with diesel. He predicts that, using this system, half of all new trucks sold in China will be electric powered by 2028.

Biofuels are the other potential solution for heavy trucks. Biomass is any organic matter, including crop residues, forest residues, crops grown specifically for energy use, organic municipal solid waste and animal wastes. Biodiesel is produced by combining ethanol with animal fat, recycled cooking fat or vegetable oil. Hydrogenated Vegetable Oil in HGVs reduces emissions by 83 percent compared to traditional diesel – although there are questions about at how large a scale such fuels could be used on and remain sustainable.

The biggest issue with biofuels is if crops are grown specifically to produce biomass feedstock. This can take land away from food production and increase pressure on resources. It is better to use waste products as the feedstock for biofuels – inedible crops, forestry products, waste oils and fats, municipal solid waste – or else use crops harvested on marginal land which does not compete with important food crops. Alternatively, using algae as biomass, which grow faster and produce energy at a rate 30 times faster than food crops, does not require land. Biofuels still produce CO_2 when they burn, but this is equal to the CO_2 absorbed by the plants and algae for photosynthesis and is known as biogenic carbon. It is part of the short-term carbon cycle and, as such, is generally excluded from carbon footprint calculations, although its impact on the atmosphere is the same. By contrast, burning fossil fuels releases carbon that has been bound up for millions of years into the atmosphere as CO_2.

Opinion in the European freight industry currently sees gas as a more realistic lower-carbon option, especially for long-distance road transport by heavy trucks. Gas-powered trucks already made up 42 percent of heavy truck sales in China in 2024, mainly due to gas being cheaper than diesel. They use liquified natural gas, but the same engines can be fuelled with biogas, essentially methane created as organic matter decomposes. Methane is created in municipal landfill waste dumps, where it presents a risk of explosions, but instead it can be captured and used as fuel. This is done either by processing organic waste in an anaerobic digester, or by covering a landfill site with earth and then drilling wells into it to pump off the gas. The same process can be used on livestock farms to convert manure and other animal waste – for example, by covering waste lagoons and pumping off captured methane.

Hydrogen fuel cells may provide another future solution: they are already a proven alternative energy source for some buses, forklifts and boats. Most hydrogen used today is produced from natural gas in a process known as steam reforming, which releases greenhouse gases – this is known as 'grey hydrogen'. 'Blue hydrogen' is produced in the same way, but the emissions are captured and stored. 'Green hydrogen', produced from the electrolysis of water using electricity from renewable sources, has no emissions and offers the promise of playing a major role in a zero-carbon freight system. Producing green hydrogen is a more expensive process, though costs are likely to come down.

At the short-haul and 'last-mile' end of the freight business, electric vehicles are a simpler proposition. Distribution and logistics companies Amazon, DHL, Fedex, UPS and DPD are increasingly using electric vans for deliveries. French-owned DPD is on track to electrifying its whole van fleet in the UK – 15,000 vehicles – by 2025. But there is still debate about just how sustainable EVs are – for instance, there is concern around what happens to the batteries at the end of their lives. Meanwhile, other EVs available to the wine industry include models of agricultural tractors and forklifts.

Rail carries eight percent of the world's passengers and seven percent of global freight transport, generating lower emissions and pollution than road transport. The rail sector is widely electrified, especially in Europe and Japan: the Swedish rail system is fully electrified and runs on 100 percent renewable energy from hydroelectric and wind-power plants. The Nordic monopolies transport wines from southern Europe by rail. Some company sites have private sidings where goods can be received or loaded directly onto rail wagons. Using rail helps alleviate road traffic – a train can take 50 truckloads – and reduces air pollution. However, rail freight is generally less electrified than passenger services. For wine producers, one option can be to use an 'inter-modal' combination of rail and road transport. Josep Maria Ribas of Torres in Catalonia reports that the company's strategy of persuading clients to take deliveries in this way has been successful to date in shipping wine to Germany and the Netherlands. He says that costs are similar, though lead times are around a day longer.

Major UK importer Liberty has been even more successful in switching to rail transport. 'We calculated that close to 40 percent of our emissions were from outsourced logistics,' says Liberty chairman David Gleave. 'Switching to train is relatively easy and it has saved us money.' He says that shipping by train from Italy, a major source of their wines, takes an extra week but 'it's just a matter of putting that into our forecasts'. Since the company committed to going carbon neutral in 2014, it has switched 90 percent of its Italian volume to train shipments, using the Verona to Zeebrugge route. The precise carbon saving by rail depends on the location of wineries: for their Soave imports, using rail cuts their emissions by 49 percent, though from western Sicily the reduction is just 25 percent. For this reason the bulk of their Sicilian imports arrive by sea in two container shipments a year. Eighty percent of Liberty's Spanish imports are now also transported by sea, from Valencia and Bilbao, which cuts their emissions by up to 60 percent.

In the US, Bronco Wine Co uses rail to ship cases of wine from California to eastern distribution partners, with one railcar replacing four big semi-trucks. And glass container manufacturer Encirc, which we looked at in Chapter 4, has a purpose-built railhead at its plant in Elton, Cheshire, which receives shipments of recycled cullet.

Lastly, air transport generates by far the highest greenhouse gas emissions. Travel for business meetings with customers, market visits and trade fairs can be an important part of doing business and building relationships. Where these are necessary, there may be alternative modes of transport with reduced greenhouse gas emissions: high-speed rail links across Europe and Asia are increasing.

One alternative mode of transport could potentially be inland waterway routes. Europe boasts 37,000km of waterways connecting hundreds of cities and industrial regions, while the US freight water highway makes up nearly 19,000km. The energy consumption of water transport per kilometre/time is estimated to be half that of rail. It is mainly used for transport of cargo such as coal, grains, ores and aggregates like sand, gravel and slag.

How can retailers green their supply chain?

Large grocery retailers import many of the wines they sell directly and have logistics teams who oversee the transport of goods from supplier to store. More broadly, their size means they have economies of scale and control of supply chains to make changes that can have a real impact. They can, as we have seen, ship wine in bulk and package in the destination market in order to reduce emissions and cost – though even for these companies, a large number of wines in their range do not sell in volumes to justify bulk shipping 24,000 litres at a time. They are also more able to offer wines in alternative packaging formats such as bag-in-box or cans, because their orders can meet minimum production runs and their rates of sale ensure wines are sold within the product life at optimum drinking quality. For wines still sold in glass bottles, several large supermarket groups have committed to the Sustainable Wine Roundtable's Bottle Weight Accord to reduce average glass bottle weights to 420g by the end of 2026. These include Tesco in the UK, US-based Whole Foods, Albert Heijn in the Netherlands and Systembolaget in Sweden. Their large-format stores can also facilitate recycling schemes for shoppers.

While the supermarkets do not own or control all the wine brands they sell, their buying power means they have a strong voice with suppliers as well as consumers. Their economic muscle can influence packaging manufacturers and logistics companies to innovate and adopt more sustainable products and practices. In this way, large grocery retailers set global standards. Also, as large companies they are increasingly required to report mandatory Environmental, Social and Governance (ESG) information annually, and to publish corporate social responsibility strategy updates, greenhouse gas emissions, and their plans to move to a circular economy. Wine producers wanting to work with big supermarkets are increasingly required to demonstrate their sustainability credentials.

Systembolaget in Sweden and all of the Nordic monopolies have been at the vanguard of wine trade efforts to improve sustainability, pushing for lighter bottles and for better labour conditions for wine workers, as well as transporting wines by rail rather than road. Admittedly, as monopolies they have an advantage of a captive market: they do not need to worry about competitors. But they have demonstrated what can be achieved, developing recycling schemes and on-shelf labelling which offer customers sustainable choices.

Systembolaget, with 450 stores, works with around 1,000 providers who supply 20,000 wines with an annual volume of 210 million litres, a quarter of which is organic. Their goal is for a 50 percent reduction in greenhouse gas emissions between 2019 and 2030, and carbon neutrality by 2045. These goals are part of a joint agreement with other Scandinavian monopolies in Norway, Iceland, Finland and the Faroe Islands. Equally important, they are open to sharing their knowledge. Systembolaget's next and most ambitious project will calculate a product carbon footprint for its whole range. 'It will be transformational,' says sustainability manager Marcus Ihre. 'We realized we needed a way to calculate our Scope 3 emissions for every product.' The footprint figure will include emissions from fertilizer, pesticides, fuel and electricity. Ihre hopes that they will tentatively be able to start sharing the figures with consumers in summer 2026.

Smaller wine companies have fewer resources and many use third-party logistics. Most of what they sell is shipped in bottle and in small volumes. They may have less direct control and less influence over their supply chain – but the principles remain the same. Building a more sustainable supply chain starts with having a clear sustainability strategy, understanding and measuring all emissions and waste:

› **First, take stock of the supply chain** and the company's role in it, then a strategy can be developed to address areas under direct control. Identify action to reduce Scope 1 and 2 emissions by moving to renewable energy, improving energy efficiency and reducing waste.

› **Understand the impact of the supply chain up and downstream.** This is important for the identification of opportunities to reduce Scope 3 emissions and understand the impact on a circular economy of all materials and goods used.

› **Introduce a Supplier Code of Conduct**, terms of trade and product specifications to reflect sustainability goals can be developed. This requires understanding suppliers and their sustainability strategies:

verifying that they are actually doing what they say they are. Anyone in a buying role can develop policies that include sustainability criteria alongside cost, quality and service considerations.

- **Bring in guidelines on acceptable bottle and packaging weights.** A switch to lighter-weight glass bottles is easy to do and collectively could make a difference to logistics-related emissions. As EPR regulations and taxes become more widespread, costs will be considerable: for companies over legal thresholds (in the UK, those with an annual turnover of more than £1 million), recording information on packaging will be necessary for submissions. Starting with own-label and fast-moving wines, where there is more control, will be easier and will have the greatest impact on emissions and cost savings.

- **Collaboration** – the open sharing of information will enable faster action. Understand the routes and transport modes available; assess which has the lowest emissions. For temperature-controlled warehouses or depot storage, question whether renewable energy sources are being used. How could energy efficiency be improved? How much waste is going to landfill?

- **Reduce and recycle packaging.** What happens to each type of packaging across the supply chain – for example, are cardboard outer cartons made from recycled cardboard? Could plastic be removed from the supply chain? Are cardboard and other recyclable materials actually recycled or do they just go to landfill? What are the collection and actual recycling rates in the area of disposal? Could they be increased, for example, by offering recycling bins or communicating with customers?

- **How can customers be helped** practically to find sustainable solutions? Could sustainable suppliers and products be better featured by providing information and insight? As we will see in the following chapter, the signs are that customers want products that help them live more sustainably.

Key lessons for a sustainable supply chain for businesses of all sizes

› **Understand the end-to-end supply chain** for all products distributed and their component parts.

› **Explore packing in the destination market** where sales volumes permit.

› **Use transport space efficiently.** Reduce weight and ship full container loads where possible.

› **Explore transport routes** and mode of transport options and evaluate your carbon footprint. Where is your product creating emissions and how could they be reduced?

› **Ask, could you reduce greenhouse gas emissions** by switching part of your wine's journey to rail? What kind of trucks is the wine being transported in by road?

› **Ensure packaging (primary, secondary and tertiary) is appropriate to the supply chain**: apply the principles of Chapter 4.

› **Operate Zero Waste to Landfill** at intermediate warehouse and retail sites.

› **Minimize energy consumption** for temperature control in transit and at warehouse sites, and use renewably generated energy.

› **Ask what happens** to your product's packaging post consumption.

› **Provide solutions for customers** to help them lead more sustainable lives – electric charging points, sustainable packaging, collection points and recycling schemes. Give transparent information to enable customers to make informed choices.

Chapter **SIX**

SOCIAL SUSTAINABILITY AND WINE

One pruning season in the South African winter, 'it was excruciatingly cold,' remembers Johan Reyneke in Stellenbosch, 'and I was wearing a surfing wetsuit to keep myself warm. And then I realized my [Black] colleagues were using newspaper under their clothes. And I thought, we've got to fix this.'

That was the late 1990s and the shivering workers weren't Reyneke's employees. His solution to such inequality, in the biodynamic wine company he founded in 2000, was to commit to buying his workers houses and to send their children to university. 'It was financially disastrous,' admits Reyneke, and remains a work in progress, though the sixth employee house will be bought this year.

Such initiatives make a real difference to workers in an industry where social justice looms larger than anywhere else in the wine world. South Africa remains a profoundly unequal country where apartheid still casts a long shadow. And as employers of some of the poorest rural South Africans, the wine industry is on the front line of dealing with those issues.

The kind of challenges faced by South African wineries and their workers over the decades following the end of apartheid might seem extreme – and perhaps very distant from Bordeaux or the Barossa. Yet while South Africa faces particularly glaring inequalities, and its wine industry has a toxic legacy to overcome, the same basic principles of fairness and business ethics apply the world over. This is what is sometimes referred to as 'social sustainability': running businesses in ways that are fair to their employees and their local communities, and which promote an economy with the broad-based strength to prosper in the long term.

People make the difference

Much of this book has focused on the environmental aspects of sustainability – yet it is people that enable an organization to function. And businesses have an impact on people too: their employees, customers, suppliers, local communities, investors and wider society. This is their social impact.

Like all businesses, wine businesses need to be profitable. They need to make money to pay employees' salaries, fund ongoing operations and invest for the future. As prominent South African winemaker Bruce Jack warns, just focusing on the environment is 'a very Northern Hemisphere view. Cutting CO_2 emissions is way down the list in Africa – we're just trying to survive.' Nora Sperling-Thiel of Stellenbosch's Delheim Wine Estate adds that for her, the most important thing is that 'we sustain a farm where 65 people make a living'.

This is the intersection between social and economic sustainability. Wine businesses need employees and they need customers. So we cannot talk about sustainability without addressing the question of what people want: understanding customers will always be at the heart of any good business.

Consumers

Wine has been enjoyed by humans for around 8,000 years; today, wines reach a wider and more diverse customer base than ever before. It's true that in every market there are a variety of consumers more or less engaged in wine. But the interest and demand for wine knowledge has sprouted new businesses, wine clubs, tasting groups, books, and specialist publications, while the Wine and Spirit Education Trust (WSET) has more people pursuing wine certifications than ever before.

Yet the irony is that global wine consumption continued to decline in 2024, following a long-term trend. The International Organisation of Vine and Wine (OIV) announced in 2025 that global wine consumption figures for 2024 were the lowest since 1961. Fewer people are drinking wine, and those who are, are drinking less. According to International Wine and Spirits Record (IWSR), the global leader in alcoholic drinks data and insights, wine volumes are forecast to continue to decline in most key markets over the next few years.

Declining wine consumption has been a feature of European markets for some time. For centuries, the biggest wine producing countries – France, Italy and Spain – were also the biggest consumers, and wine production in each was pretty much in line with domestic consumption up until the 1960s. But as lifestyles changed and populations shifted from rural to more urban living, annual wine consumption in France dropped from 113 litres per capita in the 1960s to 47 litres in 2021. The total amount of wine the French consumed dropped from 5.6 billion litres to 2.5 billion. Italy, Spain and Portugal saw similar trends.

In the last decades of the 20th century, surplus production in Europe was partly offset by developing new consumer markets. Southern European wines were exported initially to northern Europe, to countries such as the UK, Germany, Holland, Belgium and Scandinavia. And as consumption in traditional wine-producing countries continued to drop, global wine exports expanded further to the US, Japan and China. Wine consumption in China grew with the affluence of the rising Chinese middle class, peaking in 2017 at seven percent of the world's wine consumption. The proportion of global wine production exported continued to rise, from 25 percent in 2001 to 47 percent in 2021.

IWSR indicates that after rebounding post pandemic, the number of wine drinkers in key global markets declined by five million people between 2021 and 2024. Older drinkers constitute the largest age group, with over 55s accounting for more than half of wine drinkers in Belgium, France, Japan, Portugal and the UK. Previous research carried out by Wine Intelligence shows the long-term pattern across markets for young consumers; below age 34 is similarly downbeat: the proportion who drink wine at least once a month is declining steadily. In 2010, 36 percent of adults of legal drinking age in the US and around half of UK adults aged 18–34 drank wine once a month. By 2020, those numbers had fallen to 21 percent and 26 percent, respectively. We see similar trends in other key markets: wine is not appealing to younger generations in the way that cocktails and craft beers do.

One exception is Japan, where the proportion of wine drinkers under 34 increased from 10 percent in 2015 to a current level of 15 percent. And there are some other positive signs for the wine industry. The 2025 French Sowine/Dynata Barometer showed that wine is still the most popular alcoholic beverage in France. Wine also ranked first among 18–25 year-olds, for the first time. The study also found that more consumers were following wine brands or producers on social media. TikTok has become the number-one wine platform in France, overtaking Instagram: 39 percent of survey respondents said that influencers played a role in their purchase decisions. IWSR likewise sees opportunities for wine with new, younger, more adventurous consumers. According to IWSR, millennial consumers, now in their late 20s to early 40s, tend to be more highly involved in wine, have more experimental tastes and are happy to spend more on a bottle or a drink in a bar, at a time when older consumers are cutting back on spend. IWSR's CEO, Richard Halstead, says: 'Millennials who drink wine regularly report buying more expensive wines than the average for the market as a whole.'

Part of the general decline in alcohol consumption is attributed to people typically limiting their alcohol intake to lead a healthier lifestyle; an increasing proportion choose to abstain completely. Consultants McKinsey, in their State of the Consumer 2024 report, say the global wellness market is growing by five to 10 percent annually and is now worth more than $1.8 trillion. The wellness movement has driven an increase in low-alcohol and non-alcoholic beverages, while events such as Dry January have become important in some markets. But IWSR says that while moderation is driven to some degree by health concerns, it is also driven by consumers aiming to reduce their expenditure on wine – and because wine has become more expensive. Customers are feeling squeezed by increased prices, higher duty rates in the UK, and by the effect of import tariffs in the US, which have already pushed up retail prices. Further increases from the impact of EPR and increased employment costs in the UK and elsewhere will continue to increase pressure.

Most wine drinkers enjoy wine as part of a healthy lifestyle. Of course, overconsumption of alcohol is unhealthy and can cause well-documented social problems and dependency – so how wines are promoted is part of the wine trade's social responsibility. But the January 2023 WHO statement that 'no level of alcohol consumption is safe for our health' is a blunt instrument: it makes no distinction between wine, beer and spirits, or indeed between widely varying consumption patterns.

Consumed in moderation, wine has positive effects on mental health and reduces anxiety – and consumers are prioritizing mental health, with more than a third globally identifying emotional well-being as their primary health goal, according to Innova Market Insights' Top Food Trends report 2025. Socializing is good for mental health and for overall health. A wide array of medical research shows overwhelming evidence that drinking wine in moderation has positive health benefits, reducing the risks of cardiovascular disease, stroke, diabetes and some cancers – as Argentina's Dr Laura Catena, physician and fourth-generation vintner, explains on www.indefenseofwine.com. Wine is both a cultural product and part of a healthy Mediterranean diet. The French Paradox – the much-heeded observation, first made in the 1990s – attributed low levels of coronary heart disease in France, despite French people's high intake of cholesterol and saturated fat, to red wine. Gut health has also become an important focus, particularly since the pandemic: leading nutritionist and health scientist Dr Tim Spector, professor of epidemiology at King's College London, indicates that red wine has been shown to boost our microbiome.

So, marketing and communicating about wine in a responsible and balanced way is important: there are ways that businesses can help consumers manage their intake. Most countries have published guidelines for drinking alcohol, with varying recommendations. Promoting responsible drinking includes public awareness campaigns, promotional materials and labelling, along with providing spittoons at tastings. Alcohol units are a simple way of comparing different types of alcoholic drinks. In the UK, the Portman Group is the social responsibility body and regulator for alcohol labelling, packaging and promotion: its label guidelines indicate unit alcohol content per container, the pregnancy logo advising against drinking for pregnant women, signposting to drinkaware.co.uk and the UK chief medical officer's Low-Risk Drinking Guidelines. In France, Bruno Le Breton at Domaine de la Jasse in Languedoc developed a Responsible Wine Tasting Charter in conjunction with Alcohol Addiction Specialist Dr Hussan Al Mallak which was published in 2015.

And consumers – especially younger consumers – also want transparency, in particularly over ingredients in food and drink products. The quality of ingredients and of a product are important to them. There has been a shift away from meat to vegan or vegetarian diets, partly for health reasons but also due to concerns around animal welfare and meat's impact on greenhouse gas emissions. While grapes are the major ingredient in wine, additives, processing aids and preservatives may be used in winemaking, with some of these traditionally of animal origin.

There are now alternative plant-based protein options which can be used as the demand for vegan and vegetarian wines increases. Unlike food, the wine industry has not previously been required to list ingredients other than allergenic substances. But there is a need for transparency and regulations are changing. In the EU, for wine made from 2024 onwards, mandatory information includes a list of ingredients and nutritional declaration on the label or via QR code and online e-label.

IWSR research shows more than half of regular wine drinkers across key markets say they are motivated by sustainability, particularly wealthier and millennial consumers. Moreover, numerous general consumer surveys find that most consumers say they want to buy products that are good for them, for the people that make them and for the planet. PwC's Voice of the Consumer Survey 2024, addressing 20,000 consumers across 31 countries, found that 85 percent had experienced first-hand the disruptive effects of climate change in their daily lives. Forty-six percent said they are buying more sustainable products as a way of reducing their personal impact on the environment. Meanwhile, Ernst & Young (EY) Future's international consumer index for June 2022 indicated that while affordability was the consumers' first concern amid the cost-of-living crisis, consumers say they want to live in a more environmentally friendly way, especially those aged under 35. And consumers' awareness of the social justice issues bound up with the products and services they buy has increased too.

'As the wine industry adapts to climate change and is affected by more extreme weather events, wine producers are looking for ways to reduce their climate impact, and making sustainability part of their brand presents an opportunity to connect with the values of younger drinkers,' says IWSR's Richard Halstead. But, he warns, 'Gen Zs of legal drinking age and millennials are sensitive to "greenwashing" and will look for products where eco credentials are integral to a brand's identity, not an afterthought or an add-on.'

Wine quality has to remain the most important factor. Consumers don't buy Coca-Cola because it's packaged in an rPET bottle: they buy it for its taste and image. Wine consumers don't care about the exact weight of a glass bottle – whether it's 400g, 500g or 700g doesn't mean anything to them – but they do look to producers and the trade more generally to take care of all that. And some trade buyers are paying attention: certainly in the UK, sustainability credentials are increasingly a deciding factor for choosing on-trade suppliers.

We also know that many wine drinkers find wine, wine shops and the amount of choice overwhelming. Some see wine as old-fashioned – a beverage for white, middle-class men, and sometimes the way wine is portrayed reinforces that view. We also know that younger drinkers care less about traditional wine-marketing cues around, say, heritage and soil types. Yet there is also a desire among food consumers to connect with culinary heritage and embrace authenticity and tradition. There is also a trend for locally sourced produce and organic food. In wine, organic, sustainable and natural wines are gaining in appeal, while urban wineries have become a phenomenon. There are now 157 vineyards located in big cities in 35 countries. Italy boasts 32 urban vineyards in

15 cities, including seven in Venice, with others in Rome, Bolzano, Naples, Milan and Catania. And this goes hand in hand with consumers' desire for experience, connection and memories.

Millennial consumers purchase a higher than average number of grape varieties across key markets. Says the IWSR's Halstead: 'They frequent more wine-buying channels, giving extra opportunities for engagement, such as personalized experiences and storytelling to build awareness and loyalty; and local points of engagement, including pop-up events, wine dinners and wine clubs.' So with creativity, the wine industry can engage with them in a different way to make wine interesting, desirable and fun, be that through packaging formats, design or storytelling.

There are opportunities, then, to innovate and engage by developing new wines, formats, ways of communicating, shared experiences and events. There are opportunities for the industry to develop a narrative around culture, travel, socializing and the positive effects of drinking wine in moderation. Wines have stories to tell, and we can tell them in ways that reach more consumers, digitally. And by offering wines that are more sustainably produced, packaged and delivered, we can offer customers opportunities to live more sustainably. Few retail consumers might buy a wine because it is sustainable at this point in time; nevertheless, sustainability is becoming an increasingly important factor in choices made by on-trade buyers, for hotels, bars and restaurants.

Employees: working in the vineyards

Vineyard and cellar work is physically demanding – and never especially well paid. Much of the work of cultivating grapes is manual. Field workers are exposed to sun and heat in summer – temperatures can easily top 40°C in some vineyards. During the September 2023 harvest in Champagne, four workers died from heat-related health issues, including a cardiac arrest, after temperatures reached 38°C in the first week of picking. In winter, working conditions in Europe are often freezing and pruners can suffer from repetitive strain injuries. Workers' health can also be harmed by agrochemicals.

Wineries and packaging facilities are often hardly more agreeable workplaces than the vineyards. They are generally cold, damp, noisy environments with wet floors and forklift trucks zipping about. Specific hazards include the use of heavy machinery such as presses, chemicals, CO_2 and alcohol fumes, and long hours at harvest time. Yet at the same time, these are jobs that demand specialist skills: pruning vines, driving vineyard tractors and harvesting machines, identifying disease and pest risk, and other tasks all require training and special knowledge, as do working in wineries and bottling facilities.

Harvest time is especially problematic, which is something not necessarily obvious to consumers in wine-importing countries. UK sustainability expert Anne Jones says: 'In the UK we see harvest time as a cross between a village fête and a garden party.' But elsewhere, in the rest of the wine world, the vintage is a short period of intense activity, where the number of workers swells with a mix of permanent and seasonal staff. Many wine production regions increasingly rely on seasonal migrant workers to fulfil the need for more harvest staff. Burgundy, Champagne and Bordeaux depend heavily on migrant workers coming from Eastern Europe – Bulgaria, Romania and Poland – as well as Africa, South America and Spain. Meanwhile, New Zealand brings in 16,000 workers a year from Pacific countries; as does Australia, via the Recognized Seasonal Employer Scheme. New Zealand's Hunter's Wines has done this for several years: James McDonald, senior winemaker, says that conditions for the migrant workers are good, with proper accommodation and sports facilities, and wages above the minimum wage: a high proportion return to work in subsequent years.

At a bare minimum, wine employers should pay a living wage and protect their workers' welfare – whether that's protecting them from agrochemicals in the vineyard, providing pickers with rest breaks, water and sunscreen, protecting against alcohol- and CO_2-related incidents in the winery, providing ear plugs for work on the bottling line, or protecting retail staff from workplace violence. But social sustainability goes beyond this, encompassing worker concerns including pay and poverty, health and safety, job security, well-being, gender equality, and dignity at work.

South Africa: righting a history of injustice

As stories of abuse of migrant workers in the wealthiest French vineyards suggest, social sustainability is as important a goal in the First World as it is in developing nations. Nevertheless, South Africa's experience offers a particularly important example of what can happen when workers and forward-looking landowners work together to change an industry.

Wine in South Africa was historically a product of white colonialism and the racialized exploitation of the country's enslaved Black and 'coloured' (the official South African designation for mixed-race people) majority by wealthy white landowners. The first Dutch settlers arrived in the late 17th century and immediately planted vines. They brought in slaves from Asia and elsewhere to work the land: this ended only with the British abolition of slavery in 1833. Yet the division of power remained stark and deeply racialized. Many slaves stayed on their owners' land, exploited as a nominally free rural workforce.

This power imbalance worsened from 1948, when the apartheid system further institutionalized white supremacism in law. Black farm workers were ruthlessly exploited, with poverty, wages and primitive housing made worse by the infamous 'Dop' system of part-payment in wine, which led to rampant alcoholism and the prevalence of conditions such as foetal alcohol syndrome. With the formal overthrow of apartheid in 1994, vineyard workers in theory enjoyed greatly improved labour rights and a minimum wage. The South African wine industry, which for years had been backward and focused on low-quality wine, opened to investment, at the same time as export markets opened up with the lifting of boycotts on South African goods. In the 30 years since, conditions have certainly improved, but progress has been slow and uneven.

Pay is just part of the challenge. Forward-thinking producers pay wages above market rates: the rural minimum wage, in rand, from March 2025 is R5,614 (£244/US $306) a month – about £1.20/US $1.50 an hour. For example, Spier, a large producer just outside Stellenbosch, pays its field labourers R7,290/month – around a 30 percent premium.

[*continues on* page 198]

Migrant Workers

In September 2023, French authorities in Champagne shut down squalid accommodation being used by gangmasters to house 52 African grape pickers: the building had a mud floor, no ceiling, and little hot water in a bathroom authorities described as 'in a repugnant state'. The workers recounted how they were given little or inedible food and were never paid. One Malian man, named only as Mahamadou, told French daily *L'Humanité*: 'We were treated like dogs.' There were at least three cases in the same month of the authorities shutting down unsafe and unhealthy accommodation being used by gangmasters to house Ukrainians, Bulgarians and other Eastern Europeans – all in the heart of one of Europe's most prosperous wine regions, with sky-high land prices and individual wines routinely priced in excess of €50 a bottle.

Migrant workers are especially liable to sub-standard accommodation and food, long working hours, low wages, problems with legal paperwork, and the depredations of gangmasters. In 2024, Bordeaux saw a flurry of cases of human trafficking and abuse of migrant workers. In July, two Moroccan gangmasters were jailed by a Libourne court for exploiting and mistreating Moroccan migrants, who were found to have been coerced to come to France under fraudulent contracts, made to pick grapes without a break from 5am to 2pm, and housed in unsanitary conditions, for which they were docked €200 a month. In late 2024, 25 Moroccan workers, dubbed 'convicts of the Bordeaux vineyards' by the *Ouest France* regional daily, sued their employer for human trafficking, long hours, and filthy, crowded lodgings with no hot water or heating and infested with cockroaches.

Champagne is particularly heavily reliant on migrant workers, requiring around 120,000 grape pickers during the harvest. Champagne houses generally sub-contract recruitment to agencies, who, in turn, can subcontract to other employers to bring in workers – a process open to modern slavery and human trafficking conditions. These problems are long-standing. In August 2018, two Sri Lankan gangmasters were jailed after 120 grape pickers, some of them illegal immigrants, were found living in grim conditions in a dilapidated hotel in the Côte des Blancs. In that case, as in more recent ones, the growers or Champagne houses at the top of these employment chains avoided any sanction: their legal responsibility remaining unclear. Industry body the Comité Interprofessionel du Vin de Champagne (CIVC) pledged in 2023 to act, and stated that the safety of pickers is an 'absolute priority'.

Meanwhile, in the Italian Piemonte's Langhe region, home to Barolo and other prestigious appellations, local farmers' union Confagricoltura Cuneo estimates that more than half of the area's seasonal workers are migrants. In spring 2024, the authorities uncovered more than 30 cases of the '*caporalato*' gangmaster system. One Gambian man told the *Al Jazeera* news website that he was paid just €3 euros an hour. He and fellow workers lived in a small, makeshift camp in the woods with no toilets, running water or electricity. Many such migrants are undocumented, meaning they are often afraid or unable to file complaints. Matteo Ascheri, president of the Barolo producers' organization, the Consorzio Barolo Barbaresco, admitted to *Al Jazeera*: 'It's a huge problem.'

This is precisely the kind of situation companies could avoid by managing their workforce in line with sustainable values. And it is in the industry's own interest to make itself look like a viable future to the communities where wine is made.

Allan Sichel, president of Bordeaux's professional body, the Conseil Interprofessionnel du Vin de Bordeaux (CIVB) told a UK Wine Society seminar in 2025 that he thinks part of the problem is that until relatively recently, producers assumed that French labour laws protected workers: the problem was the difference between labour standards and actual labour practices. Growers too often assumed that worker welfare wasn't their responsibility if they employed contractors – yet they still have legal responsibilities in terms of working conditions, safety and the minimum wage. Suppliers and growers can now sign a charter, which Sichel says has helped raise awareness of the issues. There has also been some work on integrating the housing needs of migrant vineyard workers with those of seasonal workers on the nearby Atlantic Coast. But it is clear that Bordeaux still has some way to go to address the issue.

Elsewhere, the existence of a national minimum wage and other legal standards doesn't necessarily guarantee that workers enjoy them. In 2023, a group of 150 wine farm workers marched in Paarl to demand the proper enforcement of labour laws and highlighted poor housing and wages still below the national minimum.

Housing remains a major issue. In many places, the workforce lives on the farm as well as working there. At Bosman Wines, near Cape Town, most of the workers and their families have lived on site for generations. Bosman has now built a crèche and community learning centre in the village near their Wellington farm, as well as a retirement complex. Likewise, when Andrew Gunn bought Iona in 1997, it came with the workers and their families in a village on the property. Today, the village houses around 80 people, the extended families of 20 workers. 'We have a responsibility as farmers,' says Gunn. 'As a farmer, you're a social worker, you're the bank, you help with domestic problems,' adds his wife, Rosie.

This is why the certification scheme run by the Wine and Agricultural Ethical Trade Association (WIETA) covers not only pay and working conditions but also includes an inspection of workers' housing, conducted every three years – though they can only inspect where there is a contractual relationship regarding housing between employer and worker. WIETA was formed in 2001 out of the new traction for ethical trade after the end of the apartheid-era boycott in the 1990s. It got a boost with the development of ethical trading standards by the Scandinavian alcohol monopolies from 2009. An excoriating Human Rights Watch report on the South African industry, 'Ripe with Abuse', in 2011, was another landmark: it detailed dozens of cases of below-legal-minimum-wage workers, appalling housing, unsafe use of chemicals, and intimidation and violence by white landowners against their Black workforce.

Today, WIETA currently certifies more than 1,300 suppliers, around two-thirds of South African wine production, using third-party audit organizations and international benchmarking. They partner with NGOs, trade unions and South Africa Wine, the government trade body, and their activity has spread into areas such as training for seasonal workers on their labour rights. Certification is required by many international wine buyers: 'Any producer that is successfully exporting knows they have to do the right thing to get into foreign supermarkets,' says Cathy Brewer of Villiera Wines. WIETA chief executive Linda Lipparoni says that the uncertified producers are mostly selling to the local market – more upmarket South African store Woolworths demands certification from its suppliers, but some competitors do not – and to the rest of Africa, Russia and China. But she adds that some premium estates remain uncertified too: 'there has to be market pressure,' she says, to force many producers to do the right thing.

But the responsibility for ensuring better conditions for workers remains with producers. Briton Alex Dale, co-founder of Radford Dale, says: 'I'm very much against paternalism but there are great benefits to looking after your team.' His workers all live in the nearest village – but like other enlightened employers, he provides them with a bus to work, as well as meals. For almost 20 years, Radford Dale's Land of Hope initiative has also paid for the private education of Black employees' children and dependents: they're currently funding the schooling of 14 children.

Sophia Warner has taken a different tack. She was a London special needs teacher until she came here in 2003, aged 34. 'I just got sucked in,' she says. She founded the Pebbles educational charity that now helps operate centres serving 1,400 children and has expanded into the Cape's fruit industry. It offers health checks for pregnant mothers; a first-1,000-days programme for young children, as well as crèches and after-school clubs; mobile classrooms and libraries; social workers; clinics and dental care; and more. At a Pebbles crèche and school in Hemel-en-Aarde, taking kids of workers from nearby Creation Wines, they look after children from nine months to age 14. Children are supposed to go to high school at age 12 but transport is a big problem in a rural area like this, with the high school 20km away.

In a different vein, Spier's Tree-Preneurs scheme works with poor communities in what are euphemistically termed 'informal settlements'

– shantytowns – such as nearby Vlottenburg. Groups of people grow seedlings provided by Spier in backyards until they're 60cm-plus tall, and then barter them back for bicycles, farm tools and clothing. Where a set of school clothes for a child costs a day and half's wages – and no one in a family may work anyway – that can be significant. Spier then donates the trees – around a million to date – to local schools and other spaces.

One source of funding for such projects is Fairtrade, whose label is now well established in Britain. The buyer (for wine in the UK, usually a supermarket) pays a premium of around 90 cents (about four pence) per bottle. That then goes back to the producer to distribute to their workforce, though it cannot be given out as cash. Instead, a Fairtrade Committee made up of workers decides what products or programmes to invest in. South Africa today produces more than two-thirds of the world's Fairtrade-certified wine. At Journey's End, for example, the Fairtrade premium adds up to around R100,000 (£4,300/US $5,400) a year: the committee has in some years bought gas stoves or washing machines for the 19 permanent workers who are beneficiaries (up to 60 seasonal workers also get pro-rata benefits). At Du Toitskloof, in Rawsonville, Fairtrade premiums fund the operation of two daycare centres looking after 130 children aged three months to five years. They also offer after-school care to primary-school children, a mobile library and media centre, a clinic that sees over 200 patients a month, and four college bursaries.

And yet, despite all this, the living conditions of many people in these communities remain shocking. Most of Journey's End's permanent workers live in the nearby informal settlement of Sir Lowry's Pass. The winery funds the Mila's Angels crèche there, as well as a cooperative sewing project for women making shopping bags for sale and a soup kitchen. It also paid for improvements to the primary school. Yet the dilapidated shacks and the barbed wire on the gate of the crèche tell their own story. Female-headed households are the norm, in part a legacy of the apartheid government's policy of encouraging men to go to work in the gold and diamond mines and thereby breaking up Black families. Alcoholism, substance-fuelled domestic violence and babies born with foetal alcohol syndrome are grim facts of life. The school dropout rate at Grade 7 (age 13) is around half; youth unemployment nationally is 44 percent and must be even higher here. State welfare benefits start at around R300 (£13/$16) a month: there is essentially no safety net. Other settlements are so violent, with gang activity, drugs and protection rackets, that it is unsafe to enter them. There are some places, for instance, to which Spier cannot deliver its tree seedlings.

'You've got Third World and First World next to each other, in your face,' says Johan Reyneke. 'We have to work together.' To fix such problems will take generations. Those are the ambitious goals of South Africa's Black Economic Empowerment project. The ANC Government set a target in 2015 for 20 percent of the nation's land and water to be Black-owned by 2025: whites make up around seven percent of the population but own around three-quarters of privately owned land. Today, the actual total owned by Black South Africans – over 80 percent of the country's population of 62 million – still sits at just three percent.

'It's a difficult thing to do,' concedes South Africa Wine's Philip Bowes. 'Other professions are better at identifying talent than the wine industry, for example accountancy.' At Stellenbosch's Elsenburg Agricultural Training Institute – home of the country's foremost wine training programme – just nine percent of students are Black.

For a start, the children of farm workers desperately need better education: if anything, school standards have, in fact, slipped since the 1990s. Providers of care such as the Pebbles projects are realizing that – as at the Hemel-en-Aarde Pebbles school – provision needs to be extended throughout children's lives to have a lasting impact. That is what Bosman now try to do at their Blovei school: they realized that despite the crèche and services for younger children, the rates of those dropping out later weren't improving. With more focus on after-school clubs and the like, the numbers finishing high school have improved.

Beyond school, Bosman's Julia Moore says that vocational training provision remains poor, especially in rural areas. Yet this is what young people need in order for their lives to improve. Moore says eight former students from the school are now at the local college studying marketing, hospitality and other subjects, up from just one in 2020. The Pebbles project, based with its head office at Villiera Wines, in Stellenbosch, offers a young adult programme aimed at school leavers, working on their personal development and then work skills. It's partly a question of limited horizons in poor rural families; as Sophia Warner says: 'If your family have worked on a farm for generations, they don't know what jobs are available in the industry.' Hospitality and tourism especially are now central components of the South African wine industry. 'If they had offered a positive image of the wine industry in school, I would have done it [gone into wine],' says Berene Sauls, Black founder of Tesselaarsdal Wines.

One training programme providing young people with an introduction to such jobs is the Pinotage Youth Development Academy (PYDA), near Stellenbosch, which offers training and support to unemployed 18–25 year-olds. Wine adviser Thabang Fanana at Delheim Wines went to the PYDA aged 24 and studied wine and marketing: he did placements at wineries before coming to join the hospitality operation at Nora Sperling-Thiel's estate. Delheim also continues to support to Nitsiki Biyela, probably South Africa's highest-profile Black female winemaker, by providing facilities to make wine at their winery, where she also trained.

Efforts are also under way to create new Black wine businesses alongside successful white-owned estates. Praisy Dlamini is founder of Adama Wines, a Black business spun out from Bosman Wines. She herself comes from a family of KwaZulu-Natal sugar cane farmers. But she says she became sold on a career in wine when she saw oenology students at Stellenbosch foot-treading grapes. ('My mum said, "why would you choose this – you don't even drink wine!"') Adama's head office staff and winemaker are all female and Black and 'coloured' (mixed-race), and many of them benefitted from bursaries or went to the PYDA, including winemaker Ruth Faro. And two percent of the profits from their Amandla brand in the UK and US go into student bursaries.

Kleine Zalze, meanwhile, launched Visio vintners in 2018, a company 51 percent owned by 40 Black workers in the winery and tasting room. Kleine Zalze funded its vineyard planting and branding; now it wants the Black employee-shareholders to take control of all its operations. Visio's UK brand, The People, is now listed in Tesco supermarkets. Similarly, at Great Heart wines, spun out of Mullineux and Leeu Family Wines with that company's grapes and winemaking: 42 employees receive dividends on their shares when it makes a profit.

Yet the structural inequalities in South Africa's society are huge and deeply entrenched. Alex Dale says he came to South Africa because of Nelson Mandela: he had visited the country before and seen its impressive potential for winemaking, so he resigned from his winemaking job in Burgundy the day that Mandela was elected, in April 1994, and began a new life on the Cape. The atmosphere back then was full of hope, he recalls. However, successive ANC governments since Mandela stepped down in 1999 have proved unable to effect much change. They have also become deeply mired in corruption, especially during the presidency of Jacob Zuma, 2009–18.

In South Africa, it is hard for wine businesses doing the right thing to swim against the tide. Dale says that while some things have exceeded expectations since he came to South Africa in 1994, the ANC governments since Mandela have been a 'gut-wrenching disappointment'. And yet he, Sophia Warner, Nora Sperling-Thiel and the rest carry on trying to change the future for their workers and their children, one harvest at a time.

Wine workers beyond the vineyards

Wine workers don't just toil in vineyards and wineries. The global wine trade employs millions of people worldwide: a PwC study in 2024 found that in Europe alone, the wine trade is directly responsible for three million jobs, while the California wine trade employs 422,000 in California and 1.1 million across the US. The industry requires expertise not just in vineyards and wineries, but in exports, logistics, marketing and communication.

Closest to the consumer, most wine is sold in shops and restaurants. Jobs in retail and hospitality often involve long and anti-social hours, are often poorly paid, and can involve issues of security too. This is partly why businesses across the hospitality and retail sectors in many countries are at present struggling to fill vacant positions. Yet whereas some industries can turn to technology to replace workers or make up for staff shortages, the wine trade is based on customer experience and relationships: it needs people who can work with customers and give them informed advice.

It is ultimately a question of values – the values reflected in how things are done in an organization, its culture, relationships and behaviour. To attract staff, the wine trade needs to be able to offer decent pay and working conditions, with opportunities for training and development and career progression. As well as specialist wine expertise, the wine trade needs professionals with general management skills – leadership, finance, marketing and communications – coming from a diverse range of cultural backgrounds to reflect the diverse international consumer base.

Wine producers and rural communities

The economy of Portugal's upper Douro Valley has long depended on wine. Smallholders traditionally grew grapes on plots of a hectare or two to sell to the port houses, while others in the remote villages worked in the vineyards, especially at harvest. Yet Rob Symington, joint CEO of port producer Symington Family Wines, warns that 'there isn't a generation coming through who want to do that work'. Between 2010 and 2020, the number of grape farmers in the Douro fell by almost half.

Unfortunately, the valley's steep, rocky slopes mean the vines have to be tended manually, with little opportunity to automate vineyard tasks. Portugal's 2021 census confirmed that the decline in the rural population is ongoing: young people tend to move to the cities or the coast, or else abroad. 'We're going to get to a point soon where we see the abandonment of vineyards,' says Symington. 'The days of small vineyards are over – the input costs are just too high. There's going to be disruption – the question is whether it's planned or not.'

Symington's are trying to help manage the area's transition through a range of initiatives. They are heavily involved in discussions around regulation – local quotas for still wine grapes over those destined for port distort the market – and advocate state funds for the retirement of ageing growers. They recently put 200 freelance workers on full-time contracts to give them more security. They fund places for two agronomy students a year at the local University of Trás-os-Montes and Alto Douro, in Vila Real.

In rural wine regions throughout the world, wine businesses are at the heart of local communities. The wine trade can bring jobs and local prosperity while nurturing local identity. But in a declining market, competition is increasingly fierce and many wine companies are struggling. The economic sustainability of some major winemaking regions is in jeopardy. In Bordeaux, wine is directly or indirectly responsible for 60,000 jobs. Yet supply there outstrips demand: the French government is paying vignerons to uproot vines, and, at the end of 2024, 50 estates were up for sale. The economic crisis is compounded by the impact of climate change and extreme weather patterns. Similar issues exist in other European regions, including Rioja, where several wineries are facing bankruptcy.

Beyond their current economic difficulties, however, if wine regions such as the Douro Valley, the northern Médoc in Bordeaux, or the Barossa Valley in South Australia are going to survive in the long term, they need to be able to attract younger generations to live and work there. This is not a given in these communities the way it was a century ago: the long-term trend across all developed economies is for younger generations to move away from rural areas to urban centres.

Complaints like those of Symington are now common across much of the winemaking world. In Chile, 'working in the vineyards is hot, with long hours – it's hard work,' says Viña Emiliana's head of sustainability, Sebastián Tramón. 'The workforce is getting older and getting sick. There's a real challenge in bringing in young people.' He adds that it isn't helped by poor local infrastructure in rural areas, for instance access to healthcare. At Doña Paula, in Argentina's Mendoza region, many of the fieldworkers are Bolivian immigrants: since few graduated from high school, the winery provides them with an hour a day of schooling at a school on site in Luján de Cuyo. And in northern Greece, near the remote Macedonian town of Siatista, Dimitris Diamantis of Magoutes Vineyard reports similar challenges. 'It's difficult to hire workers for this hard work,' he says: most wine workers here are over 40 and Albanian. 'We want to be able to give a motive for young people to work here,' he adds, but says that without higher wine prices, higher wages are unsustainable.

Allan Sichel says that even in Bordeaux, the industry struggles to attract workers because of the low cultural status of agricultural work. The city's Cité du Vin exhibition says: 'Take a class of third years [UK: Year 10; US: 9th grade – 14–15 years old] in a Gironde school, ask the pupils to raise their hand if they know someone who works in the wine sector, and everyone will raise their hand. Now ask them how many of them see themselves as a cellarmaster, or tractor driver, or vigneron: no hands go up or hardly any.' Sichel emphasizes the need for career evolution for workers and getting them involved in training others. There are also efforts to reach out to local schools: a group of six top producers have launched the *L'appel de la Vigne* (call of the vine) initiative, training workers to do presentations on their work and careers to local schools.

The role of wine tourism

The sun is struggling to come out on an overcast April day at Bolney Wine Estate in Sussex, in the UK, but events manager Kelly Markwick is upbeat. Standing among rows of vines on the gentle slope facing Bolney's new visitor centre, she is in discussion with vineyard manager, Phil Grant, about plans to replant this plot with the Champagne-specific clones Chardonnay, Pinot Noir and Pinot Meunier: 'How do we make this vineyard as pretty as we can?' asks Grant.

It's a question for many English producers who have embraced tourism as part of their businesses. Bolney, England's fifth-biggest wine producer, was bought by international sparkling wine giant Henkell Freixenet in 2022. This has given it access to serious investment – culminating in the £1 million (US $1.25 million) visitor centre with restaurant, tasting room and shop. Bolney welcomes around 30,000 visitors a year. 'Locals are your bread and butter,' says Markwick. But she is keen to talk about visitors from further afield too: Bolney is just 20 minutes' drive from Gatwick airport: 'Global visitors are important in terms of exports, for getting people engaged with the brand.'

It seems a long way from the equivalent scene at Boschendal, in Stellenbosch. At this historic winery, amid colonial buildings and huge trees, families picnic in the shade, ravishing views of Simonsberg mountain in the distance. Boschendal claims to have been the most Instagrammed vineyard in the world in 2023 and gets up to 350,000 visitors a year. Meanwhile, at Spier, just outside Stellenbosch town, you can sprawl in the sun with your bottle of wine on an immaculate lawn, or take part in tastings, eat at a choice of restaurants, or stay in the luxury hotel on site.

It might be more exotic than Sussex, but the business logic for wineries is similar. In Chile, wineries like those at Montes and Santa Rita, and in Argentina at Zuccardi and Doña Paula, boast stylish restaurants: in both countries, Brazilians are the largest group of wine tourists, with daily flights from São Paulo to Mendoza. Meanwhile, at Quintessa, in California's Napa Valley, staff host tastings in small pavilions dotted around the estate, looking out over stunning landscapes. At Gundlach Bundshuh in Sonoma, the vibe is a little more rustic, though artfully so, with visitors gathering for tastings around wooden benches.

Indeed, the wine tourism model is at its most developed in California: the state's wineries have more than 25 million wine-related tourist visits annually. But while wine tourism started in earnest in Napa in the mid-1970s, California can really thank the movie *Sideways* (2004) for making wine tasting cool. The film's road trip of wine geek and Pinot Noir obsessive Miles and his friend Jack through the Santa Barbara wine country helped establish visiting wineries as a fun thing for non-wine trade people to do. Its effect is well attested by many Californian producers. But wine tourism is a key part of the burgeoning English wine trade too. For smaller producers, cellar-door sales can amount to a large proportion of their turnover, enabling them to sell direct to consumers without discounting to middlemen or stores.

Rathfinny, just outside Alfriston in Sussex, is a more ambitious operation. Owner Mark Driver made his fortune in the City, spending at least £10 million (US $12.5 million) to establish the 242-hectare wine estate with an ambition of making more than one million bottles of sparkling wine a year (they're currently at around the 300,000 mark). Tourism is an integral part of the business, with a restaurant, picnics, bed and breakfast and self-catering accommodation, and a flourishing wedding venue operation. English producers have taken their cue not just from California but from New Zealand and parts of Australia too. Leeuwin Estate, in Australia's Margaret River, features a concert series as well as high-end dining options. At d'Arenberg, in McLaren Vale, there's not only a fancy restaurant but also a museum, the futuristic Cube tasting room, and the opportunity to blend and name your own wine at the Blending Bench.

The New World and English model of wine tourism is less evident in the wine-producing heart of Europe. While travellers might see plenty of *dégustation* signs in places like the Loire and the Rhône, they mostly indicate just that – tastings of a producer's wines. These tended to be stop-off points for families heading from Paris to holiday homes down south, so they could collect wine on the way for the summer. Apart from the big Champagne houses in Reims and Epernay, few French wineries have traditionally had restaurants, though these are becoming increasingly common. Commanderie de Peyrassol, an organic estate in Provence that dates back to the 13th century, has a Michelin-starred restaurant and separate bistro; visitors can see its art collection and follow paths around the vineyard and forest.

According to Ludovic Walbaum, head of wine tourism at French tourist development agency Atout France, wine tourism was worth €518 million in 2023 to independent producers: 'It's a high value-added activity,' he told *Le Monde* in 2025. Gavin Quinney, British owner of Château Bauduc, cautions that the approach in Bordeaux is typically mixed: 'Some embrace it, lots do a half-baked job, and some avoid it like the plague.' Partly, he says, this is because at the top end, Bordeaux wines are all sold through *négociants* (merchant middlemen) and *courtiers* (brokers) and they're not trying to make sales to visitors. Some actively avoid cellar-door sales: the attitude, he says, is 'we don't get our hands dirty with that kind of involvement – it's not really the done thing'. But some high-end châteaux do take tourism more seriously, such as Château Troplong Mondot, in Saint-Emilion, which offers tours, high-end accommodation and its Les Belles Perdrix restaurant. Bordeaux opened the Cité du Vin in 2016, a foundation for wine culture and civilizations housed in a contemporary, purpose-designed building, and voted by National Geographic as among the top museums of the world.

Meanwhile, Rioja has long encouraged and received visitors from across Spain and further afield. Wine producer Vivanco has a Museum of Wine Culture and many Rioja houses attract visitors through building and design, including the Frank Gehry-designed iconic hotel and spa at Marqués de Riscal, the wavy roof at Bodegas Ysios by Calatrava, or Zaha Hadid's winery for Bodegas López de Heredia. The Antinori winery in Tuscany, south of Florence, opened in 2012, designed to welcome wine enthusiasts. Built into the hillside and entirely from local materials, it was the first time in 26 generations that Antinori had inaugurated a new winery. It was awarded the World's Best Vineyard in 2022, the same prize given to Marqués de Riscal in 2024.

Another European region starting to offer a more complete visitor experience is Portugal's port country. In Vila Nova da Gaia, traditional base of most of the port producers across the mouth of the Douro from Porto, there's now a labyrinthine World of Wine Experience, featuring visits, museums, wine dinners and cultural events such as *fado* shows. Big producers there, including Taylors's and Symington's, owners of Grahams, offer restaurants and well-appointed tasting rooms as well as tours. But Rob Symington, co-CEO of his family firm, sees a more central role for tourism than that in the grape-growing heartlands of the Douro. This spectacular highland area is being hit by a combination of problems familiar across Europe and beyond: ageing growers tending small plots and labour shortages as young people leave for the cities and the coast. 'We do believe there's a future for the region, where there's a mix between tourism, higher living standards, and better jobs,' says Symington. 'We intend to be part of that journey.'

It isn't simple, however. High-end tourism demands a skilled labour force, with hospitality training and language skills. In agricultural areas like the Douro, that workforce barely exists. So the challenge is to create career paths in the new industry, for instance by investing in staff training – as well as by offering staff accommodation. Yet at Symington's main Douro Valley site of Quinta do Bonfim, in Pinhão, with two restaurants and a hotel, one eatery was unable to open for two years because of staff shortages, despite offering competitive salaries. The company also funds a post for an officer charged with helping local businesses get going in sustainable food production and tourism.

Wine tourism also offers an example of the trickiness of measuring carbon footprints: should wineries include the transport emissions of their visitors as part of their Scope 3 emissions? Jason Haas at Tablas Creek thinks they should, though it's a difficult source to reduce, especially since most wineries are in rural areas accessible only by car. Tablas Creek's 26,000 annual visitors are thus its third-largest source of carbon emissions. At New Zealand's Felton Road, owner Nigel Greening says they add tourists to their Scope 3 emissions figure – indeed, he says: 'We add on anyone who has to visit the winery for any reason.'

Wine businesses beyond the vineyards

Investors and financial institutions are increasingly interested in environmental, social and corporate governance (ESG) issues: the need to report on them will exert increasing pressure on wine-selling businesses to investigate their portfolio and push producers for more sustainable products. We have outlined the impact of such efforts in greening wine's supply chain – but there is also work to be done on the social sustainability of wine businesses.

In February 2025, Areni Global, an independent think tank dedicated to the future of fine wine, shared research into skills shortages in the wine trade following a series of interviews with members of the global trade and recruiters. What they found was that people loved working in the wine trade. They value the deep relationships, the wonderful products and the great food and hospitality. Wine is a complex and specialized area and many wine professionals have a high level of product knowledge.

However, at times, the focus on wine is to the exclusion of general business skills. The research identified 10 missing skill sets in the industry:

1. Financial acumen
2. Project management, planning and organizational ability
3. Sales strategies and techniques
4. Consumer and market research
5. Content creation
6. Data analysis
7. Resilience, flexibility and humility
8. Negotiation and problem solving
9. Strategic thinking, vision and initiative
10. Editing and writing

It also noted that wine 'experts' sometimes talk down to customers; while they may have lots of product knowledge, the ability to create a content strategy and a media strategy, and professionally execute it, was lacking. We would argue that an understanding of sustainability is another skill set that is important for the future.

There is also a growing recognition of the need for diversity and inclusion within the wine industry as consumer demographics change and market dynamics evolve. As consumers look to support businesses that align with their values and beliefs, companies that demonstrate a commitment to ethical practices, social responsibility, and diversity and inclusion will improve their reputation and brand image.

There are still relatively few non-white faces in the European wine industry. In 2020, the Diversity in Wine survey in the UK found that just over two percent of the wine trade employees surveyed identified as Black and just over three percent as Asian. Many respondents said that discrimination in the industry is rife: the report concluded that the UK trade has 'significant' problems with ethnic diversity. Magnavai 'Mags' Janjo, a Black British wine merchant originally from Cameroon, established BAME Wine Professionals in 2020 with the goal of highlighting Black, Asian and minority ethnic talent in the industry. Meanwhile, in South Africa, the government's Black Economic Empowerment initiative has supported Black business. The Cape Wine Guild, an association of top South African winemakers, has, since 2006, run a Protégé Programme aimed at advancing Black winemakers in the industry, offering a three-year paid internship with Guild members.

As for gender diversity, a UK Women in Wine Survey in 2023 revealed that 78 percent of respondents felt that sexism, gender bias and harassment were serious issues; 44 percent had considered leaving the industry as a result. In 2020, America's prestigious Court of Master Sommeliers was rocked by a scandal when multiple women accused leading male figures in the Court of persistent sexual harassment. But an American Association of Wine Economists working paper in July 2024 on The Gender Equality Landscape found that the number of women in leadership positions in the Australian wine industry had improved at all levels over the last decade, particularly in the number of women CEOs. However, it also found women's presence was still low, with just 21 percent of Australian viticulturists and 16 percent of winemakers being women. The paper also noted the disparity in pay, with women earning significantly less than male counterparts. In France, 28 percent of vine growers are female – though this is higher than the 20 percent of women in other forms of French agriculture. So there is a way to go – in part a reflection of wider society. But a diverse and inclusive workforce gives a wider range of perspectives, generates more ideas, is more creative, and attracts more talent.

In order to to protect their own brand and reputation, wine importers and retailers should understand the social impact of the wines they buy, the social responsibility of their suppliers, and know where there are potential risks or issues, in order to ensure minimum legal standards are achieved. In 2021, for example, the Swedish alcohol monopoly, Systembolaget, commissioned a Human Rights Impact Assessment from Oxfam on working conditions in the Italian wine industry, The Workers Behind Sweden's Italian Wine. It looked at Tuscany, Piemonte, Puglia and Sicily and found low wages and exploitation of migrant workers were common. The point of the exercise wasn't to justify a boycott of Italian wines, but such a prominent international buyer taking such detailed interest in conditions in this way could help ratchet up pressure on wine producers who are ignoring their social responsibilities. The Nordic monopolies have played a similar role in pushing for certified worker rights in South Africa. Systembolaget has also engaged with Champagne's CIVC in the wake of the 2023 harvest deaths, using its leverage with the other Nordics: sustainability manager Marcus Ihre says they have had a 'very good response'.

Greenwashing

Finally, it matters how wine businesses communicate all these efforts. There may be strong incentives for wine companies to talk up their sustainability credentials. But that can be deeply problematic when those credentials aren't very good. This is what is known as 'greenwashing' – companies hyping their record to make it sound more impressive than it is. They should be able to substantiate any claims they make – or else risk damage to their reputation.

For products made or sold in the EU, consumers should soon have better protection from greenwashing in the shape of the 'Green Claims Directive' proposed in 2023 and now making its way through the EU's legislative process. The proposed directive would require companies to substantiate any green claims they make in business-to-consumer communications, by essentially proving their claims, for instance through life-cycle analysis. It will, for instance, prohibit climate claims based on carbon offsetting. EU inter-institutional negotiations began in January 2025 and a final agreement on the directive was expected later that year.

Meanwhile, wine businesses who want to tell their sustainability story without slipping into greenwashing need to be honest and be accurate. UK watchdog the Advertising Standards Authority cautions that businesses should 'focus on near-term progress rather than long-term or aspirational goals'. A big part of effective sustainability communications is telling a good story: the complexities of making sustainable wine are a central part of the story of the farmers who strive for it. A real commitment to sustainability isn't just a bolt-on: it flows through everything a business does, alongside the financial bottom line.

Much of the problem stems from the fact that there is no legal definition for the terms 'sustainable', 'eco-friendly', or 'environmentally friendly', let alone 'green' or 'natural'. Claims that a particular packaging format or particular method of farming is the most sustainable are misleading.

There are numerous 'sustainable' certifications for wine. Each one has its own value and offers consumers some degree of reassurance. But each is different, with its own set of rules and measures. Widely recognized organic and biodynamic certification schemes, as well as the Fairtrade label, do not on their own constitute sustainability. How do we weigh up an organically produced wine that in theory is better for the environment but which may have a higher carbon footprint, or adds to toxicity in the soil with copper, or which is packaged in a heavy glass bottle? And as far as we are aware, none of the certifications take into account distribution issues. We return to these challenges in our conclusion.

Key principles for social sustainability

- **Know your customers**: know their needs now and anticipate their needs in the future.
- **Make sure workplaces meet legal health and safety requirements.**
- **Understand the concerns** of direct employees and also of your supply chain.

- **Create a diverse team** and an inclusive work environment.
- **Encourage, support and develop** employees, so they can realize their full potential.
- **Make sustainability a part of your business culture** – encourage sustainable business practices: your team can come up with new and creative ways to be more sustainable.
- **Offer schemes to help employees lead more sustainable lives** – electric car schemes, charging points, bicycle purchase, sustainability training.
- **Think about how your business fits** into the local economy. For example, if you are a wine producer, are there any opportunities to expand your business as well as create new kinds of jobs through tourism and hospitality?
- **Be honest about your products**: don't be tempted to 'greenwash'.
- **Create opportunities to share information** with your customers and end consumers in person, in store and through media – for example, on working sustainably and packaging.

CONCLUSION

THE WAY FORWARD FOR WINE

From a warm field in Essex to frost in Chablis, we started this book by surveying some of the extreme weather events that wine producers are now having to contend with. Wine is the canary in the coal mine of climate change: proof that the climate emergency is already changing our world and some of the most fundamental assumptions about one of humanity's longest-established agricultural products. On the land, sustainability is literally a question of ensuring that people can continue to make wine in Bordeaux or Chile or South Australia in 50 or 100 years' time. It's no longer a given. That is this book's first message: climate change is here already, and it's an existential threat to wine and to the people who make their living from it.

Over the course of this book, we have shown how wine businesses – and especially winemakers – are responding to that challenge. But even though it is the biggest and most complex problem currently facing humanity, sustainability is not simply about climate change. It encompasses a range of other issues too: the use of chemicals, the health of our soil, and the well-being of the people working in the global wine industry and the communities it sustains. So our aim has been to set out what is meant by sustainability – and what it means for wine. We wanted to increase our understanding of the environmental, social and economic aspects of sustainability – and to share examples of wine businesses that are working positively on these challenges. That is this book's second message: sustainability is about more than the environment, and that it touches every stage of an everyday product's journey from the field to our dinner tables.

Third, any effort to make wine more sustainable must start with a better understanding of that journey. That is true for any product:

especially, but not exclusively, industrial products, like the bottles that wine comes in. If we really want to live more sustainably, we need to understand not just where the glass of wine in our hand came from, but also the origins of the table it stands on, the story of the bowl of olives we are enjoying with it, the phone we glance at as we sip. We have followed wine's journey and all of its implications for sustainability: as a society, we need to do that for everything.

And as we have seen, this is no simple task: sustainability is complicated. There is no one sustainable wine, no one farming method, no one supply chain that applies to all wine or is always the most sustainable option. It all depends where and how grapes were grown, how the wine was made, how and where it was packaged, how it is sold, and what happens to the packaging after we've drunk it. And at times there are conflicts between different aspects of sustainability. Is shipping wine in bulk to reduce its carbon footprint the right thing to do if it takes away jobs from poorer parts of the world? Is buying wine in bag-in-box, with a better carbon footprint than glass but using an unrecyclable metallized plastic bag, the more sustainable choice? We are accustomed as consumers in an advanced capitalist society to being able to make simple choices that fix these sorts of dilemmas. But there just isn't a simple answer – or a simple choice for consumers who want to choose sustainable wine.

So fourth, we have to do what we can and make imperfect choices. That is precisely what the wine producers we've seen in these pages are doing – and indeed the best of them provide a model for other kinds of businesses trying to understand and reduce their environmental impact. Sustainability, in wine, as in anything else, is a journey and direction of travel, not a box-ticking exercise. You can always do better. But just because there isn't a neat, complete solution, that doesn't mean it isn't worth striving to make wine more sustainably. And, indeed, while the challenges are huge, as we have shown, there are inspiring examples of good practice around the world – from the vineyard to the winery, from the factories that make bottles to the ships that transport them, and from wine shops to consumers' tables.

As for the wine trade, we also posed some questions at the start of this book about the other challenges that wine businesses are currently grappling with globally, including falling demand and other economic difficulties. Our fifth key message is that sustainability is inevitably intertwined with such challenges. We are often asked by wine businesses whether adopting sustainable practices will allow them to increase their prices. Given the current economic struggle for many of them, that's a fair question. The answer is maybe not: there is no easy guarantee of sustainability and consumers are also feeling the squeeze. But while there are costs associated with certain sustainability initiatives, there are also cost savings. Meanwhile, sustainable methods bring increased resilience in dealing with climate change. And adopting a sustainable approach will open up more channels to market for wine producers, as sustainably produced wine becomes the expected norm for trade buyers and, increasingly, retail customers.

As Richard Halstead of IWSR says: 'Making sustainability part of their brand presents an opportunity to connect with the values of younger drinkers.' Wine needs to work on new ways of engaging with consumers in all markets through creative storytelling, creating experiences, and developing new products and packaging formats. And if the wine trade makes itself more diverse, it will find new and more creative solutions.

As we have seen, the global wine trade is complex and fragmented. We have tried to include examples that reflect how companies and people are addressing issues in different places and parts of the supply chain. There are many others – we could not include them all. But there are some consistent themes as we move from vineyard to glass and beyond. The first is the need to switch to renewable energy sources wherever possible, as soon as possible. The second is to minimize energy consumption and thirdly, to create circularity – in the vineyard, in the winery and with wine packaging across the supply chain. That is easier said than done: wine businesses will need a high degree of collaboration, particularly in addressing issues such as the supply chain and Scope 3 emissions.

The challenge for consumers is not simple either. There are numerous 'sustainable' certifications for wine. Each one has its own value and offers consumers some degree of reassurance. But each one is different, with its own set of rules, and needs to be considered in context. Widely recognized organic and biodynamic certification schemes, as well as the Fairtrade label, do not on their own constitute sustainability. And, as far as we are aware, none of the certifications take into account distribution or packaging: these crucial aspects of sustainability are invisible on wine labels. So we have to be honest with consumers and transparent about what goes into wine.

For those who want more information on, say, suitability for vegans, additives, sustainable practices or glass bottle weights, it should be available. This is where wine sellers can help consumers make informed choices.

Yet all of us who drink wine – not just the companies that make and sell it – can be part of the way forward. We all have choices about the products we buy and where we choose to buy them. To drink wine more sustainably, we can:

› **Choose organic, biodynamic and regenerative wine**, from grapes farmed with a consience. Those aren't answers to sustainability – but they're a start.

› **Support winemakers in your region or country**, if wine is made there. The wine has to travel less far to your table, and you'll be supporting local economies.

› **Choose not to buy wines in heavyweight glass bottles**, and avoid screw-cap wines – or at least put the cap back on the bottle in the right recycling bin.

› **Recycle all wine packaging.** Wash containers and separate waste into the right bins.

› **Encourage retailers to introduce collection points**: for cork and flexible plastic packaging. And encourage local restaurants and bars to recycle glass bottles and investigate the use of returnable bottles.

- **Know what happens to packaging** in your city or area. Look for packaging made from recycled materials that are readily recycled in your area, and challenge local administrations to improve collection schemes and increase actual recycle rates.
- **Share your concerns with winemakers and retailers**: ask why they have used the packaging they have.
- **Be open to new ideas** and alternative packaging formats, particularly for everyday drinking wines.
- **Look at the sustainability credentials** of wine producers and the companies you purchase from. Be challenging, and look at the detail. Information is not always available – but the more we ask for it, the more likely it is that it will be provided.

But just as important: enjoy your wine! It remains one of humanity's most sublime inventions: let's ensure that future generations can carry on enjoying it as we do.

APPENDIX 1:
BIBLIOGRAPHY

Links and resources that we've used in this book, by chapter:

Introduction

BBC Reith lectures – Mark Carney 2020:
https://www.youtube.com/watch?v=uvw-aC0KLD4

'Climate Change – Observations, Projections, and General Implications for Viticulture and Wine Production', by Gregory V Jones (paper presented at Climate and Viticulture Congress, Zaragoza April 2007: https://www.infowine.com/wp-content/uploads/2024/04/libretto4594-01-1.pdf)

Climate, Grapes and Wine – vintage reports and climate appraisals from Gregory V Jones: https://www.climateofwine.com/publications

Ellen Macarthur Foundation – 'What is a Circular Economy?': https://www.ellenmacarthurfoundation.org/topics/circular-economy-introduction/overview#:~:text=Circular%20economy%20explained,remanufacture%2C%20recycling%2C%20and%20composting.

European Environment Agency – Economic Losses from Weather- and Climate-Related Extremes in Europe: https://www.eea.europa.eu/en/analysis/indicators/economic-losses-from-climate-related – October 2024

EY Future's International Consumer Index 'In Crisis, but in Control': https://www.ey.com/en_gl/insights/consumer-products/future-consumer-index-in-crisis-but-in-control – June 2022

Fairtrade – 'Wine':
https://www.fairtrade.net/en/products/Fairtrade_products/wine.html

INRAE (Institut National de la Recherche Agronomique), March 2024:
https://www.inrae.fr/en/news/global-map-how-climate-change-changing-winegrowing-regions

OIV (Organisation Internationale de la Vigne et du Vin), World Wine Production Outlook, November 2024:
https://www.oiv.int/sites/default/files/2024-12/OIV_2024_World_Wine_Production_Outlook.pdf

PwC's Voice of the Consumer Survey 2024:
https://cee.pwc.com/voice-of-the-consumer-survey-2024.html

The Future We Choose – Surviving the Climate Crisis (2020) by Christiana Figueres and Tom Rivett-Carnac.

TedX Greta Thunberg: https://www.youtube.com/watch?v=EAmmUIEsN9A

United Nations Sustainable Development Goals – 'Take Action for the Sustainable Development Goals': https://www.un.org/sustainabledevelopment/sustainable-development-goals/

United Nations Sustainable Development Goals – 'The Sustainable Development Agenda':
https://www.un.org/sustainabledevelopment/development-agenda/

BK Wine Magazine – 'How important is wine to the country economy' by Per Karlsson, April 2022:
https://www.bkwine.com/features/more/important-wine-country-economy/#:~:text=Globally%2C%20world%20wine%20exports%20measures,as%200.04%25%20of%20global%20GDP

World Meteorological Organization – State of the Gobal Climate 2024: https://wmo.int/publication-series/state-of-global-climate-2024

Chapter 1 – Climate Change

A Life on Our Planet – *My Witness Statement and a Vision for the Future* by David Attenborough (Ebury Press) 2020.

Château Cheval Blanc – Manifesto promoting (anti)conventional viticulture: https://www.chateau-cheval-blanc.com/wp-content/uploads/2024/03/EN-manifeste-page-a-page.pdf

Climate Action Tracker: https://climateactiontracker.org/

European Environment Agency – Economic Losses from Weather- and Climate-Related Extremes in Europe, October 2024: https://www.eea.europa.eu/en/analysis/indicators/economic-losses-from-climate-related

IEA (**International Energy Agency**) – Global Energy Review – CO_2 Emissions: https://www.iea.org/reports/global-energy-review-2025/co2-emissions

INRAE (**Institut National de la Recherche Agronomique**): Climate Change – European Vineyards Most Affected, May 2025: https://www.inrae.fr/en/news/climate-change-european-vineyards-most-affected

IPCC (**Intergovernmental Panel on Climate Change**) Special Report – Global Warming of 1.5°C: https://www.ipcc.ch/sr15/

IWCA (**International Wineries for Climate Action**) – Taking Collective Action to Decarbonize the Global Wine Industry: https://www.iwcawine.org/

Monarch Electric Tractors – 'Ready for Vineyard Operations': https://www.monarchtractor.com/vineyard-tractor

La Revue du Vin de France – '6 % des vignerons veulent quitter le bio dans les 5 ans' by Jérôme Baudouin:
https://www.larvf.com/6-des-vignerons-veulent-quitter-le-bio-dans-les-5-ans,4839511.asp

Sabi Agri Electric Tractors, 'Empowering Agroecology with Smart Agricultural Equipment':
https://sabi-agri.com/en/

Soletrac Electric Tractors, 'The Best Tractor for a Vineyard':
https://solectrac.store/tractor-for-vineyard/

Sunagri (pioneering dynamic agrivoltaics): https://sunagri.fr/en/

The Switch – *How Solar, Storage and New Tech Means Cheap Power for All*, Chris Goodall (Profile Books) 2016. *And see* Carbon Commentry:
https://www.carboncommentary.com/

United Nations Climate Change – The Paris Agreement:
https://unfccc.int/process-and-meetings/the-paris-agreement

United Nations Environment Programme – Plastic Pollution:
https://www.unep.org/plastic-pollution

World Meteorological Organization – State of the Gobal Climate 2024:
https://wmo.int/publication-series/state-of-global-climate-2024

Worldometer – World Population By Year:
https://www.worldometers.info/world-population/world-population-by-year/#google_vignette

Chapter 2 – Growing Grapes

OIV (Organisation Internationale de la Vigne et du Vin) – State of the World Vine and Wine Sector in 2024:
https://www.oiv.int/sites/default/files/2025-04/OIV-State_of_the_World_Vine-and-Wine-Sector-in-2024.pdf

Chapter 4 – Packaging

Sustainable Wine Round Table – Reducing Bottle Weight:
https://swroundtable.org/wp-content/uploads/2024/10/SWR-Report-Reducing-Wine-Bottle-Weight-2023.pdf – June 2023

The Porto Protocol – Unpacking Wine Guide:
https://www.portoprotocol.com/product/unpacking-wine-guide/

Chapter 6 – Social Sustainability

Areni Global – 'Rethinking Education: Shaping the Future of the Wine Trade':
https://areni.global/rethinking-education-shaping-the-future-of-the-wine-trade-2/

IWSR (International Wine and Spirits Record) – What's Driving Wine's Structural Decline?
https://www.theiwsr.com/insight/whats-driving-wines-structural-decline/ – July 2024

Responsible Wine Tasting Charter:
https://cloud.maison-le-breton.com/s/ZnI1.qLJk7DrZwBgxBSb

APPENDIX 2:

GOVERNING AND CERTIFYING BODIES

This list features both governing bodies and certification agencies that feature on these pages and others that are recognized in terms of wine sustainability assurance.

AAWE (American Association of Wine Economists) – a non-profit educational organization based in New York City.
ACO (Australian Certified Organic) – Australia's leading organic and biodynamic certifier.
AEMO (The Australian Energy Market Operator) – manages electricity and gas systems and markets across Australia.
AGW (A Greener World) – promotes verified farmers and food outlets that deliver positive impacts for the global environment.
Areni Global – an independent think tank dedicated to the future of fine wine.
BAME Wine Professionals – highlighting Black, Asian and minority ethnic talent in the wine industry.
BDA (Biodynamic Association) – UK body responsible for organic and biodynamic certification within the UK.
BEE (Black Economic Empowerment) – South African governmental policy aiming to facilitate broader Black participation in the economy.
Biodyvin (Syndicat International des Vignerons en Culture Bio-Dynamique) – European biodynamic group of 225 wine growers who farm biodynamically.
BioGro – New Zealand's largest certifier of organic produce and products.
CCOF (California Certified Organic Farmers) – non-profit organization advancing organic agriculture.

CIVC (Comité Interprofessionel du vin de Champagne) – trade association representing Champagne growers and merchants.
Close the Glass Loop – a European partnership dedicated to increasing the quantity and quality of recycled glass.
COP (Conference of the Parties) – decision-making body of the UN Convention on Climate Change, which meets every year.
Demeter – Demeter Biodynamic Certification is used in over 65 countries to verify that biodynamic products meet international standards of production and processing.
DFA of California (Dried Fruit Association of California) – a non-profit organization supporting fruit and other agricultural producers.
Directive 2009/128/EC – EU Directive on Sustainable Use of Pesticides (aims to establish a framework for reducing the risks and impacts of pesticide use on human health and the environment).
ECOCERT – Europe-based certification body offering a range of international certifications in organics and sustainability.
Ellen MacArthur Foundation – a UK-based international charity focused on accelerating the transition to a circular economy.
Elsenburg Agricultural Training Institute – South Africa's first centre for agricultural training (in Stellenbosch).
EPR (Extended Producer Responsibility) – shifting responsibility from local authorities and taxpayers to businesses for the end-of-life management of their products, including collection, recycling, disposal.
Fairtrade – works with farming co-operatives, businesses and governments to make trade fairer.
FIVS (Fédération Internationale des Vins et Spiritueux) – an international federation working for the overall sustainability of the global alcoholic beverage sector.
FSC (Forest Stewardship Council) – international non-profit agency safeguarding the world's forests, tackling deforestation and biodiversity.
GHG Protocol (Greenhouse Gas Protocol) – the most widely used greenhouse gas accounting standards; providing a framework for businesses and governments to measure and report greenhouse gas emissions.

Green Claims Directive – aiming to prevent unfounded claims of green credentials (to prevent greenwashing).

HVE (Haute Valeur Environnementale) – French certification showcasing commitment to protecting and enriching the environment.

IEA (The International Energy Agency) – providing analysis, data and policy around real-world sustainability solutions for all.

IMO (International Maritime Organization) – global agency responsible for the safety and security of shipping and preventing pollution.

INRAE (Institut National de la Recherche Agronomique) – France's national research institute for agriculture.

International Tin Association – dedicated to supporting the tin industry and expanding tin use.

IPCC (Intergovernmental Panel on Climate Change) – United Nations body for assessing the science related to climate change.

ISO (International Organisation for Standardization) – non-governmental international organization publishing international standards covering technology, management and manufacturing.

ITF (International Transport Forum) – an intergovernmental organization with 69 member countries; acts as a think tank for transport policy and is the only global body that covers all transport modes.

IWC (International Wine Challenge) – rigorous and impartial global wine competition (London based).

IWCA (International Wineries for Climate Action) – a collaboration of environmentally conscious wineries dedicated to reducing carbon emissions in the wine industry.

IWSR (International Wine and Spirits Record) – the global leader in drinks marketing analysis.

La Place de Bordeaux – global marketplace for prestigious Bordeaux and other high-worth wines.

LCA (Life Cycle Assessment) – a method for the environmental assessment of products and services, covering their life cycle from raw material extraction to waste treatment.

LEED (Leadership in Energy Efficiency and Design) – globally recognized green rating system promoting sustainable building practices.
LETIS – South American-based third-party certification body (ISO accredited).
NOP (National Organic Program) – US-based accreditation programme ensuring organizations meet national organic standards.
OIV (Organisation Internationale de la Vigne et du Vin) – the intergovernmental organization dealing with technical and scientific aspects of viticulture and winemaking.
PPWR (EU Packaging and Packaging Waste Regulation) – rules governing packaging that aim to reduce waste, improve recyclability and promote the use of recycled content.
PYDA (Pinotage Youth Development Academy) – equipping unemployed youth with the skills and personal development to work in the wine and other industries.
Regenified US based recognition for farms, ranches, forests and producers committed to the growth of regenerative agriculture.
The Regenerative Viticultural Alliance (RVA) – a verification and certification programme promoting practices that restore ecosystems and regenerate soil, verified by ECOCERT.
Respekt-Biodyvin – An association of 26 wineries from Austria, Germany, Italy and Hungary aiming to produce top-quality biodynamic wines.
ROC (Regenerative Organic Certified) – certification scheme for food, textiles and personal care ingredients, to meet high standards in soil health, animal welfare and farm worker fairness.
Rodale Institute – US-based (including Pennsylvania, California, Iowa and Italy) agricultural research and education non-profit organization dedicated to advancing regenerative organic agriculture.
SAWIA (South Australian Wine Industry Association) – the body representing viticultural and winemaking interests of the state.
SIP (Sustainablity in Practice) – California-based sustainability certification programme for winegrowers and winemakers.

SWNZ (Sustainable Winegrowing New Zealand) – certification programme for applicable industry-wide for New Zealand wine.
SWR (Sustainable Wine Roundtable) – independent global platform dedicated to advancing sustainability in the wine industry through collaboration.
SWSA (Sustainable Wines South Africa) – non-profit organization promoting sustainable practices within the South African wine trade.
Systembolaget – the government-owned chain of liquor stores in Sweden (this monopoly is the only retail outlet permitted to sell beverages containing more than 3.5% alcohol in this country).
Terra Vitis – France-based certification for sustainable viticulture.
The Porto Protocol – global movement sharing knowledge to empower the wine industry to take action to mitigate climate change and promote sustainability.
UN (United Nations) – assembly of the world's nations to discuss common problems and find shared solutions that benefit all humanity (193 member states).
USDA Organic – US Department of Agriculture labelling for agricultural products produced according to standards set by the NOP.
WIETA (Wine and Agricultural Ethical Trade Association) – a South African non-profit voluntary association of many stakeholders who are committed to the promotion of ethical trade.
WineGB (Wines of Great Britain) – the national association for the English and Welsh wine industry: leads and supports sustainable growth in GB wine.
WSET (Wine & Spirit Education Trust) – world leader in drinks education.

INDEX

A

B

C

L

M

ABOUT THE AUTHORS

Jane Masters MW

Jane Masters MW has spent her professional working life in wine in a range of senior management roles across winemaking, retail buying, marketing and business consultancy. After an honours degree in Biological Chemistry, she trained as a winemaker at the Institut d'Oenologie in Bordeaux, France, and became a Master of Wine in 1997 (she is one of 422 in the world, and was elected chairman of the Institute of Masters of Wine 2016–18). Her wine experience is complemented by an MBA from London Business School.

Jane's initial interest in sustainability stemmed from a desire to understand climate change. She quickly found that climate change, environmental, social and economic sustainability cannot be considered in isolation. 'As my understanding increased, I became aware of my own personal impact and the need for my behaviour to change. We can all make a difference at work and at home: the starting point is understanding.' Jane is a founding member of the Institute of Masters of Wine Sustainability Committee, and, since 2021, has moderated a successful series of sustainability webinars.

Dr Andrew Neather

Andrew Neather is a freelance British journalist. A former academic historian, environmental campaigner, political speechwriter and newspaper journalist, he was the *London Evening Standard*'s wine critic 2005–15. He now blogs weekly on wine and food at https://aviewfrommytable.substack.com/, writes a regular column for Tim Atkin MW's website, and has contributed to publications including *The Independent*, *Harpers Wine & Spirit* and *Club Oenologique*. He lives in South London.

ACKNOWLEDGEMENTS

Jane Masters MW

My thanks go to all the friends, colleagues and wine professionals around the world who I have worked with, and come into contact with, over my decades'-long career. The experience and knowledge they have generously shared has contributed immeasurably to the content of this book. They know who they are!

Over the years, I have also taken inspiration from other sources, and I would particularly like to thank Christiana Figueres, ex-UN executive secretary for climate change, who was instrumental to the historic COP Paris Agreement in 2015 and co-author of *The Future We Choose* (2020); Chris Goodall, whose book *The Switch* (2016) showed the move to renewable energy to be a realistic possibility; Mark Carney, ex-governor of the Bank of England and now prime minister of Canada, for his BBC Reith lectures; Sir David Attenborough, for his passion for the natural world and relentless campaigning; and, finally, Greta Thunberg, whose first TEDx event recorded in Stockholm when she was an unknown 15-year-old, still sends a shiver down my spine. It is not only what these individuals have said and written, but also their unwavering hope, sheer determination for action, and resilience in the face of doubt, that has motivated me to add my voice in this book.

My thanks also go to my co-author, Andy Neather, our editor Susan Keevil and publisher Hermione Ireland. Finally to my friends and family, in particular my partner, Rob O'Neil, for all the support and positive encouragement he has given.

Dr Andrew Neather

Many people are to be thanked for their help in the creation of this book, providing time, advice, contacts and organizational support – but especially the winemakers, who took the time to speak to me, welcomed me, and in many cases fed and accommodated me on the road. There are simply too many people to name, but my special thanks go to Derek Mossman Knapp, Maricruz Antolin, Iro Koliakoudakis and Markus Stolz. Both Tim Atkin MW and Dom de Ville at the Wine Society gave me online space and encouragement to develop some of the ideas in these pages. Thanks go to Amanda Selby for introducing me to Hermione Ireland, the Academie du Vin Library's managing director, and to our editor at AdVL, Susan Keevil. More important still, it has been a real pleasure working with Jane Masters MW: I've gained a friend as well as a colleague. Last but not least, my wife, Zillah Watson, has been eternally supportive and, as ever, my best sounding board.

A special note of thanks

To everyone who supported *Rooted in Change* before it even reached the shelves: your pre-orders and early encouragement have been greatly appreciated. Your interest has helped us launch this book into the world, and we are grateful for your support from the very start of its journey. Particular thanks go to Hill Smith Estate, Majestic Wine, Familia Torres, Chene Bleu, González Byass, Hartenberg Estate, Lallemand, Spier and Taylor's Port.

ABOUT THE PRINTING OF THIS BOOK

We have tried to produce as environmentally friendly a book as possible. To this end we are printing locally in the United Kingdom and the United States of America to have books as close to our main markets as possible. We have used vegetable oil based inks and digital inkjet toner and can confirm the book is recyclable in a paper waste stream. This book has been printed using FSC™ certified paper in line with our continuing commitment to ethical business practices, sustainability and the environment.

We have looked at the surface lamination process and decided to use a standard matt laminate because it is recyclable, it is removed as part of the de-inking process and is then used as waste for energy. With biodegradable laminate there is a concern on micro plastics going into the food stream as it only degrades in the right circumstances. Therefore recycling is a far better option than incorrect degrading.